T0236451

SpringerBriefs in Applied Sciences and Technology

PoliMI SpringerBriefs

More information about this series at http://www.springer.com/series/11159
http://www.polimi.it

Claudio Martani

Risk Management in Architectural Design

Control of Uncertainty over Building
Use and Maintenance

POLITECNICO
DI MILANO

Springer

Claudio Martani
Department of Architecture
University of Cambridge
Cambridge
UK

ISSN 2282-2577   ISSN 2282-2585   (electronic)
ISBN 978-3-319-07448-1 ISBN 978-3-319-07449-8   (eBook)
DOI 10.1007/978-3-319-07449-8

Library of Congress Control Number: 2014947127

Springer Cham Heidelberg New York Dordrecht London

Printed on acid-free paper

Springer is part of Springer Science+Business Media (www.springer.com)

The best way to predict future is to shape it.

Peter F. Drucker

Preface

The work presented in this publication is divided into six chapters organized in three parts. The first part introduces the theoretical framework.

Chapter 1 aims to underline the opportunity of controlling the uncertainty over important long-term objectives in architectural design by adopting risk management methods and techniques. With reference to the problems of maintenance that can arise from the design, the chapter outlines the contribution that the risk management process and techniques can offer in supporting clients and designers during both the decision moments of the brief and the design phases (in analysing the context, simulating events, anticipating possible scenarios), as well as in the operation and management phase (in monitoring performances of buildings in use in order to instruct actions to manage risks over time).

Chapter 2 aims to present the risk management process as a means to control risks over building use and maintenance, by estimating the uncertainty in the most important objectives. To this purpose the framework offered by the International Standards in creating a common ground to deal with uncertainty is analyzed and examples of applications of the risk management process in various contexts are reported. Finally, the possibility to use such an approach in architectural design to control risks over building use and maintenance is put forward.

Chapter 3 aims to propose the use of process monitoring on buildings to gain continuous feedback from their use, in order to monitor risks and learn by use. Both types of building monitoring approaches are introduced in this chapter: punctual monitoring, to periodically check the level of performance of buildings, and continuous monitoring, to allow timely response to changing conditions. Three examples of real-time continuous monitoring and responding systems are reported from projects of the MIT | Senseable City Laboratory.

The second part introduces a methodological experimentation.

Chapter 4 aims to propose a set of tools and methods to manage the risks over a number of objectives over the building process. The final outcome of the process proposed is a dashboard (tableau de bord) in which a level of risk is reported for all requirements. To that end, a set of tools and methods are also introduced in the chapter, with the scope to evaluate both the importance and the uncertainty over all requirements, by correlating the need for maintenance and the level of maintainability of all elements involved with each requirement.

Chapter 5 aims to propose a process for managing, at the design stage, the risks over long-term objectives. The proposed process consists of two steps: risk assessment, to assign a level of risk to all requirements on the dashboard, by attributing to each of them a degree of importance as well as of uncertainty; and risk monitoring and review, to continuously check the actual performance of buildings, and to compare them with those expected in order to update in real-time databases and evaluations.

In the third part results of applications are presented and discussed. Chapter 6 aims to test in real-world case studies the set of tools and methods to create a dashboard. Tests were run on two buildings of worship. To this purpose a degree of importance was assigned to a set of requirements that represent the needs of the Italian Council of Bishops (CEI) with reference to the phase of use of churches. Then, a degree of importance was estimated for all requirements by using spreadsheets and Monte Carlo simulations. The main results are presented and commented upon.

Acknowledgments

Foremost, I would like to express my sincere gratitude to my advisor, professor Cinzia Talamo, for the continuous and patient support during my Ph.D. Her role was fundamental in both the construction of the present thesis and the education of its author. For the same reason I would also like to acknowledge professors Giancarlo Paganin and Claudio Molinari, co-advisors with professor Talamo of the work.

Sincere thanks go to the Italian Council of Bishops (CEI), in particular the national agency for buildings of worship (*Servizio Nazionale per l'Edilizia di Culto*) and his chief Giuseppe Mons. Russo, for the enthusiastic cooperation, and for providing case studies to run applications. With them I would also like to thank the staff of the ecclesiastic local agencies (*Curie*) where applications were conducted, as well as the architects of the churches. Without their kind support I could not have tested the "dashboard". Many thanks also to Fulvio for his generous contribution.

Due acknowledgment goes to the structures in which I had the pleasure to work and study; namely the B.E.S.T. (Building Environment Science and Technology) department and the Ph.D school of the Politecnico di Milano.

My gratitude goes also to the Massachusetts Institute of Technology, and specifically to the Senseable City Laboratory, for having me as a research assistant in 2011. In particular, I want to acknowledge the Director of the Laboratory—Prof. Carlo Ratti—and Prof. Rex Britter for the year of high education and vibrant cultural activities that has been for me working and learning close to them.

I would like to thank the Virtual Development and Training Centre (VDTC) of the Fraunhofer Insititute (IFF) of Magdeburg as well, for giving me the opportunity of an outstanding summer internship in 2009. The experience gained with them was of high value in the subsequent activities of my Ph.D., which then resulted in the present work.

Many thanks also to colleagues from the University of Cambridge and to Katelin Hansen for the precious support in editing and proofreading the text.

Finally, I would like to express my most heartfelt thanks to all people who, even if not directly involved with the research, always play a fundamental role in providing me with joy and peace, fundamental conditions for working with passion. In this regard, a special sense of thankfulness goes to Martina, who to me is the force that *move il sole e l'altre stelle*. Many thanks also to my parents and all dear friends.

Grazie

Introduction

Within the building process the phase of operation and management is often characterized by a number of problems whose origin lies during the project design.

These problems often result in reduced levels of performance and the resulting discomfort for occupants due to accelerated degradation, which requires intervention—sometimes very considerable—to provide solutions. Indeed, interventions to solve problems whose origin lies in the design are likely both to change the identity of the original project and to result in high costs of intervention.

Most of the problems that cause accelerated degradation stem from the fact that there is in the current practice of the building process a poor ability to relate the design choices with the requirements for the phase of operation and maintenance— even though it is the design itself that largely determined the satisfaction of these requirements.

The building process seems to be characterized by a perimeter for phases of the objectives and responsibilities. Phases may stand alone and this means that some operators may focus on specific objectives of one of the phases (e.g., control of costs of construction) instead of on those of the whole process (e.g., control of the life cycle costs). If it is understandable that autonomous decisions in each phase are desirable for operators with specific objectives (e.g., the construction company), less understandable is the fact that it is also accepted by operators with interests in the whole process (i.e., the client).

Considering this scenario from the point of view of management it can be argued that in a series of moments of the building process a number of choices are made that have major repercussions on the operation and management phase, from the planning of intervention, to the architectural configuration of the project, until the final design. The decisions taken in each of these moments, if left unchecked, can have a domino effect—and sometimes even in a synergistic way—to the management activities.

Moreover, it has to be considered that maintenance and management have an important impact, not only on performances, but also on the costs of buildings throughout their lifetime. Therefore, there is a significant interest, both technological and economical, to improve the control of design on buildings adequacy to be efficiently maintained over time.

With reference to the recognition of difficulties in the current design process to control the effects of decisions over the operation and management phase, a plausible hypothesis is to adopt the contribution from studies on risk management, which have the ability to analyze the context, simulate events, and anticipate possible scenarios, and offer tools for support to decision.

With reference to supporting decisions, the risk management process can bring an important contribution to the building sector because of the ability to use feedback information from operation and management to understand and manage errors, either by translating feedback from previous experiences into design specifications for future design (predictive risk management) or by using live data gathering for real-time responding (risk management through process monitoring and responding).

At the present time in the construction industry, methods for risk management are still applied only in a timely and utilitarian way, and only to segmented phases and for limited subjects; for example in the context of due diligence process for acquisitions, insurance policy definition, preparation of fire safety plans, or for project validation.

In light of the difficulties during the key moments of design to envision impacts of decisions on the long-term objectives, there seems to be scope for a new application of the risk management process and techniques to architectural design, for managing risks on objectives for the phase of operation and maintenance.

The aim of this work is to create tools and methods to manage, while designing, the risks over building operation and maintenance objectives. Starting from the definition of a framework of objectives, special attention is paid to the relationship between the design choices and maintenance activities.

Contents

Part III   Applications: Tests on Case Studies

Part I
Theoretical Framework: The Management of Risks Related to Operation and Maintenance Objectives in Architectural Design

Chapter 1
The Risks of Decisions with Long-Term Impacts Within the Building Process. The Uncertainty in Design Over a Set of Objectives for the Operation and Maintenance Phase

Abstract The aim of this chapter is to underline the opportunity to control the uncertainty over important long-term objectives in architectural design by adopting risk management methods and techniques. Satisfaction with long-term objectives for buildings largely depends on the possibility to carry out due maintenance and the building process is characterized by a sequence of phases where decisions that strongly define adequacy of buildings to be properly maintained are taken at the early stages. Despite the key role of the brief and the design phases in defining building quality over time, the current practice of architectural design is characterized by a large uncertainty around the propensity of designed buildings to meet long-term objectives. This uncertainty is due to a difficulty in linking design features to the needs of use and maintenance. The operation and management phase is, then, the time after the design from which consequences of decision taken upstream appear. At this stage quality of buildings is visible and measurable and, therefore, feedback can help to track the origin of problems. This process of learning from past experiences is called learning by using. With reference to the set of problems that can arise from design the risk management process and techniques can support clients and designers by helping during both the design phase (in analysing the context, simulating events, anticipating possible scenarios) and the operation and management phase (in monitoring the performance of buildings in use in order to instruct actions to manage risks over time).

Keywords Uncertainty · Risk · Design · Brief · Maintenance · Building process · Learning by using

C. Martani, *Risk Management in Architectural Design*,
PoliMI SpringerBriefs, DOI 10.1007/978-3-319-07449-8_1

1.1 Uncertainty Over a Framework of Objectives in Architectural Design: The Risks of Failure in Goals Achievement

Within the building process the phase of operation and management is often characterized by a number of problems whose origin lies in the project design. These problems often result in reduced levels of performance and result in discomfort for occupants due to accelerated degradation and high costs of management. In this sense it is significant to consider that, according to Perret (1995), the costs for management (that includes maintenance, running costs, demolitions and restoration) can reach up to 75–80 % of the global cost of a construction (in a 50–60 year lifecycle), against 2–4 % of pre-design activities, 2 % of project design and the 15–20 % of construction. Problems in maintaining the most recent built environment are leading nowadays to increasing pressure on the topic of design for maintenance. Buildings are items designed to meet a set of needs over decades and their performance during the operation and maintenance phase is largely determined at the early moments of conception—the brief and design phases (Ciribini 1983). The rest of their useful life is then a long period of time in which the inertial outcomes of the early decisions appear. Therefore mistakes in the early stages, which lead to difficulty in maintenance and usage, can have a significant impact in terms of both performances and running costs throughout the building's lifetime. In light of this it can be considered that implications of design choices on the operation and maintenance activities are crucial both from a technological and by an economical point of view and that, for this reason, it is particularly appropriate to manage, right at the design stage, the risks related to the adequacy of building performances to meet over time promoters' and users' needs.

The difficulty in estimating, while designing, the propensity of a building to maintain adequate performances over time lies on the fact that this attitude depends on two aspects that are largely unknown in the current practice of architectural design: the amount of intervention that elements will need and their level of maintainability. Indeed, the more a building is characterized by a degree of maintainability adequate to carry out the due maintenance, the less its performance over time is uncertain (Molinari 2002). In particular with reference to maintainability, design factors[1] are particularly crucial, since they involve aspects such as: simplification of structure, facilitation technical understanding, improvement of

---

[1] As design factors are considered all the dimensional, distributional, morphological, technical and performance characteristics, inherent in the design of a building, one of its subsystem or one of its components, able to affect the maintainability. An indicative list, even if neither exhaustive nor homogeneous, of the design factors is as follows: level of complexity of the entity, typological characteristics, distribution and general geometry of the building, level of modularity of the building and its subsystems, accessibility (internal and external) of the building and its subsystems, location, size, organization and ergonomics of the operating spaces, visibility of the building and its subsystems, configuration of the components, portability of components, modularity of components, standardization of components, disassemblability and reassemblability of components, flexibility of

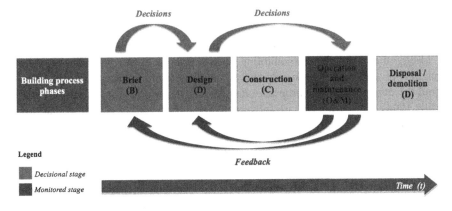

Fig. 1.1 Decisions and outcomes of the decisions during the building process

reliability, facilitation of interventions and the reduction—or at least control—of maintenance time.

The lack of confidence of designers in the adequacy of maintainability of their proposals bring as a consequence that the long-term condition of a designed building is a result not logically deducible from the available information (De Finetti 2006) at the early stage of the process. In other terms, the propensity of buildings to meet through time a set of given goals is strongly uncertain when decisions are taken.

The notion of uncertainty is referred, by definition, to those events that are impossible or difficult to predict (Gigerenzer 2002), and it can be distinguished in two types: the random uncertainty—which depends on the characteristics of the system and can generally be traced back to random variations, fluctuations and occasional stochastic phenomena; and the epistemic uncertainty—that depends, instead, on defects of knowledge and perception of the analyst (Zio 2012). Unlike aleatory uncertainty, the epistemic one can be not only estimated, but also treated and reduced (De Finetti 1990). Epistemic uncertainty is often treated using statistics. Indeed, when it is not possible, or not convenient, to deduce expected outputs of a process given all inputs, a reasonable solutions is to adopt the frequencies of outputs recorded in the past as probability of the outcomes of other processes in similar conditions. Regarding building design it is not possible to accumulate over time a statistical database of results to be used for this purpose, because buildings are not standardized items and therefore previous experiences very often cannot improve the level of confidence on future outcomes. Nevertheless, even though all buildings are quite unique, their elements and components are more standardized. Therefore, feedback from use and maintenance of elements and components (Fig. 1.1) are very

(Footnote 1 continued)
the system, clarity and definition of functional performance levels of components, availability of a maintenance plan in advance. *Source* Molinari (2002).

useful, if related to maintainability (Paganin 2005), because they reveal the quality of the decisions taken in design and bring precious indications for both improving future design and supporting the preparation of new brief documents (Talamo 2010).

Ideally the feedback or monitoring system should ensure that the need for undertaking data-retrieval (monitoring) is notified to those responsible so as to avoid the defective, or inappropriate, items being incorporated into future designs (Mills 1994). The feedbacks are fully capitalize only from organizations that promote several building initiatives over time and that, therefore, are able to use previous experiences to enhance future interventions. Examples of these organizations are big real estate owners and promoters, such as banks, insurance companies, ministries, religious congregations, etc.

The current separation between decisional and operational moments in the practices of building process is made clear by the fact that nowadays there are neither tools nor methods to track the cause-effects relationship between design choices and the availability of elements over time (i.e. it is not possible within the current design practice to predict the propensity of a building, or its parts, to be still able to guarantee the required functions within 5 or 10 years from the implementation).

In other words despite the centrality of informed design in defining the future performance of buildings there is, however, still a lack of tools and methods to take into account uncertainty over requirements (both technological and user requirements), while designing.

The need for such tools and methods has become always more urgent as the building construction process has evolved and new conditions have arisen that are required to manage uncertainty in the long-term. In this sense some relevant topics that have characterized the recent evolution of the building sector are: new actor involvement (i.e. consultants), integrated procurement routes, the use of output specifications, the growth of green buildings, the adoption of life-cycle costing techniques.

The hypothesis of this work is to propose a set of tools and methods to manage, right at the design stage, the risks of inadequate performance over time by evaluating both maintainability and exposure to failure of all technological elements involved in the design, in order to improve the consciousness of promoters and designers on the uncertainty of technical requirements. In particular to support the assessment of maintainability a significant contribution is provided by the studies on the material culture (*cultura materica*). For more complex requirements, such as user requirements, a possible solution that is put forward in this work is to estimate uncertainty by using the Monte Carlo method, which allows to estimate the outcomes of multi-variable processes, passing through simulations instead of through analytical computation (Marseguerra and Zio 2002). This method is particularly reliable because it avoids the widespread mistake—named "flaw of average"—to use average inputs for estimating most probable outcomes of processes (De neufville and Scholtes 2011). Instead it uses probability distributions as input conditions for simulations.

ability of organizations to capitalize on feedback from the operation and
ment phase relies on whether those organizations are permanent rather
ry. Even if in constructions industries promoters are generally temporary
ganization (Sinopoli 1997), examples can be found of permanent organi-
hat promote and manage several building initiatives. In particular in Italy
nt example in this sense are:

s and insurance companies, that own and manage several offices
dwide.
try of Army, which owns and manages several barracks, through the army
ctorate for infrastructure.
try of Health, which owns and manages many districts and hospitals to the
"ASL".
talian Church, which owns and manages an important patrimony over the
national territory.

reference to organizations like these learning from the use in order to
future processes of design and management are particularly valuable.
earch for a "fil rouge" between design, construction and problems in
ent would require a purely diagnostic and in-depth analysis, whose
y is mainly due to the fact that it involves actions and phenomena that
very different times from each other. Where the causal link is between
d objects, which are physically and temporally separated, is always
d and confused.
er difficulty lies in the necessity to reconstruct the overall "space con-
' in which failure events occur, where such a configuration may change
ly with changes in techno-typological characteristics of each building, its
sage profiles, the environment in which it is settle.[2]
st of an analysis of this nature is often so high as to be completely
le. It can be of great help then to use the historic "memory" of these
row 1962).
ectiveness of such a knowledge-management tool on the one hand is
proportional to the amplitude of the time horizon in which this "memory"
vely consolidated in integrated flows of information. On the other hand,
ioned by the ability to filter the data from the "memory" of the people
ystematic framework of information that is organized and easy acces-
nberg 1982). The framework of information can only be built through a
collection of information during the operation and management phase
erly collected and organized, helps to form a base of knowledge useful
and inform briefs and designs of future projects (Talamo 2003).

---

through the application of methods "Failure Mode and Effects Analysis" (FM
de Effects and Criticality Analysis" (FMECA) and the subsequent construc
t trees" (or "errors trees").

1.2 Brief, Design and Operation-Management Phases. Relations Between Design Choices and Quality Over the Building's Lifetime

The building process is characterized by a sequence of phases where decisions are taken at the early steps of brief and design. The decisions taken in these crucial moments go to influence, with a cascade effect, building performance during the operation and management phase. Despite the strong links between brief, design and quality in usage, however, in the current practice of building design there is little vision from decision makers on the long-term consequences of their choices.

In the logic of the approach for processes (UNI EN ISO 9000:2005) uncertainty is largely determined by interface between the various processes. In particular this is true for items (usually not only just the material or the elements but also of information) that are generated by the previous step of the process and become input elements for the following ones.

Indeed it can be considered a particularly critical element the fact that the uncertainty associated with the information that is generated by a process is, in a certain sense, amplified when information affected by uncertainty serves as an input for the following process; this progressive amplification of uncertainty may generate in the process as a whole a level of uncertainty particularly important but at the same time difficult to predict and manage (Villemeur 1992).

In essence, if the input data of a process are highly uncertain it is plausible that, by taking decisions upon these information, in the following steps such uncertainty is further amplified. A single fault—in, for example, the definition of a unit price that must be associated with a quantity of machining particularly important—may result in a large error in the evaluation of the intervention as a whole (Cerutti and Paganin 2012).

The information that is generated or processed in different sub-processes are of different nature and complexity, and each may have a degree of uncertainty and variability that can influence, positively or negatively, the results of the construction process.

Considering this scenario from the point of view of management it can be said that in a series of moments choices are made that have major repercussions on the operation and maintenance phase, from the planning of the intervention, to the choice of architectural configuration, to the final design. The decisions taken in each of these moments go to influence, with a cascade effect, building behaviour during the operation and management phase and, if left unchecked, with no control on the final result.

In these crucial early phases of the process the actors that have the main influence and control over the process, are the client and designers, which therefore have also the major responsibility for the final building behaviour, while those in charge of running buildings' operation and carrying out the due maintenance have a sense of the quality of design in usage.

With reference to this scenario there is a significant interest in looking at the three phases that play the most prominent role in determining the satisfaction of the promoter's needs throughout the building process: the brief, the design and the operation and maintenance.

(a) The Role of the Brief Phase with Reference to the Satisfaction of the Promoter Needs Throughout the Building Process

The brief document is a document prepared by the head of the procedure. It contains technical and administrative in-depth specifications for the design and information such as the situation pre-intervention, the general objectives to be pursued and the strategies to achieve them, the needs to be satisfied, the rules and technical standards to be met, the specific constraints of the law linked to the specific context in which the intervention is provided, the functions that the intervention must fulfil, the technical requirements that must be met, the impact of the intervention on the environment, the phases of the design process to be developed and their logical sequence, as well as the relative times of execution, the levels of the design elaborated both graphic and descriptive to be prepared, the budget limits to be observed, the estimated costs and sources of financing to be used (Maestosi 2004).

In final analysis, the brief document is the input that instructs the design (which is the most prominent and crucial decisional action of the process). Therefore, any lack of clarity or error in preparing the brief is a source of uncertainty on the future phases of design, construction and use.

(b) The Role of Design with Reference to the Satisfaction of the Promoter's Needs Throughout the Building Process

The input information required for the design phase are: brief document and programming outcomes, while the output are: design drawings (at the different levels of detail) and contract framework.

The design process is highly exposed to risks. Project design risks evolve from misunderstanding project objectives and threats to achieving these objectives. This is particularly important for aspects relating to design brief, whole life objectives and parameters, design programme, whole life cycle cost (WLC) plan, time constraints and quality issues. The success of understanding and translating project whole life objectives into a reality depends on the design team's in-depth knowledge of the whole life-cycle process and associated sources of risk. This situation is further complicated by the fact that design processes are highly interactive and involve an extensive exchange of information between design stakeholders (Preiser and Vischer 2005).

In final analysis then, what is decided in the design is nothing less than what will ideally be realized and then used. It should be the translation in a physical output of the needs expressed by the clients in the brief, and it is required to satisfy that set of needs throughout the whole building lifecycle.

The ability of designers to envision effects of their decisions in the phase of operation and management, and to design according to that, determine the level of risks of design solutions over the satisfaction of clients' need in the long term. Design is without doubt the key decisional moment of the whole building process

and the vast majority of problems that can occur in carr maintenance can be tracked back to design errors.

(c) The Role of the Operation and Maintenance Ph Satisfaction of the Promoter Needs Throughout the

Operation and management phase is one of the parts concerns the greater length of time within a building l

Operation and maintenance is the phase in the moment of design when the inertial outcomes of ear reason, in this phase feedback can be gathered to re taken upstream.

Programming tasks and expectations for the mana considered the moments of connection between the operational phase of the building process.

The opportunity to consider the parameters of tainability and substitutability, overall cost, in additi from the changed conditions in the construction m rational use of existing resources and optimization construction life cycle (Molinari 1999).

The ease, time and costs of carrying out due mai adequacy of the maintainability of the building and the phase of operation and maintenance can be c which the quality of design in use can be evaluate

In conclusion, if designers properly consider maintenance, then the quality of buildings durir maintenance will be totally satisfactory, otherwise from high cost of maintenance to discomfort in gathered from the phase of operation and maintena design with reference to long-term objectives.

1.3 Learning by Using. The Role of F to Improve the Effectiveness of Br Managing the Risks Over the Lor

As mentioned in the previous chapters, the oper time after the decisional steps from which upstream appear. At this stage quality of buildi therefore, feedback can be gained and reported t be defined as learning by using. Learning by manent building developers that can take advan in the future.

In the final analysis, the collection of data from the phase of operation and management serves to verify the quality of decisions taken upstream and to point to what could have been done better. This process of learning by using is useful for several figures within the building process, in particular for:

- Designers, in order to understand what set of choices causes most of the problems in use and in maintenance.
- Clients, in order to understand what important aspects haven't been both specified in the brief and verified in validation.

Parties that have had the possibility to see the impact of their mistakes and to track their origin are more likely than others to not repeat the same mistake again. With reference to this aspect it should be considered that designers are professionals of the building process and, therefore, they are likely to repeat the process many times. Clients instead can be of many different types but, as mentioned in this chapter, there are numerous continuous promoters that can fruitfully learn from past experience to do better in future.

References

K.J. Arrow, The economic implications of learning by doing. Rev. Econ. Stud. **29**, 155–173 (1962)

A. Cerutti, G. Paganin, *Risk management per l'edilizia* (Dario Flaccovio editore, Palermo, 2012)

G. Ciribini, Durabilità e problemi manutentivi nelle attività di recupero, in: Recuperare n.6, a.II, luglio-agosto (1983)

B. De Finetti, *L'invenzione della verità* (Raffaello Cortina editore, Milano, 2006)

B. De Finetti, *Theory of Probability. A Critical Introductory Treatment* (Wiley, Chichester, 1990)

R. De neufville, S. Scholtes, *Flexibility in Engineering Design* (MIT Press, Cambridge, 2011)

G. Gigerenzer, *Reckoning with Risk: Learning to Live with Uncertainty* (Penguin Books, London, 2002)

P.C. Maestosi, *La manutenzione nel processo edilizio* (Alinea editrice, Firenze, 2004)

M. Marseguerra, E. Zio, *Basic of Monte Carlo Method with application to System Reliability* (LiLo-Le Verlag, Hagen, 2002)

E. Mills, *Building Maintenance & Preservation. A Guide to Design and Management* (Butterworth Heinemann, Oxford, 1994)

C. Molinari, in *Il nuovo quadro di riferimento tecnico-normativo per la manutenzione*, ed. by Curcio Manutenzione dei patrimoni immobiliari. Modelli, strumenti e servizi innovativi (Maggioli, Rimini, 1999)

C. Molinari, *Procedimenti e metodi per la manutenzione edilizia* (Esselibri editore, Napoli, 2002)

G. Paganin, *L'acquisizione delle informazioni per la manutenzione dei patrimoni immobiliari* (Sistemi Editoriali, Napoli, 2005)

N. Rosenberg, *Inside the Black Box: Technology and Economics* (Cambridge University Press, New York, 1982)

N. Sinopoli, *La tecnologia invisibile. Il processo di produzione dell'architettura e le sue regie* (Franco Angeli, Milano, 1997)

C. Talamo, *Il sistema informativo immobiliare* (Sistemi editore, Napoli, 2003)

C. Talamo, *Procedimenti e metodi della manutenzione edilizia*, vol. 2 (Sistemi Editoriali, Napoli, 2010)

A. Villemeur, *Reliability, Availability, Maintainability and Safety Assessment*, vol. 1 (Wiley, Chichester, 1992)

E. Zio, *Notes from the Seminar of Mathematics Culture* (Politecnico di Milano, Milano, 2012)

Standards and Laws

International Organization for Standardization, 2005. UNI EN ISO 9000:2005 Quality Management Systems. Fundamentals and vocabulary

1.2 Brief, Design and Operation-Management Phases. Relations Between Design Choices and Quality Over the Building's Lifetime

The building process is characterized by a sequence of phases where decisions are taken at the early steps of brief and design. The decisions taken in these crucial moments go to influence, with a cascade effect, building performance during the operation and management phase. Despite the strong links between brief, design and quality in usage, however, in the current practice of building design there is little vision from decision makers on the long-term consequences of their choices.

In the logic of the approach for processes (UNI EN ISO 9000:2005) uncertainty is largely determined by interface between the various processes. In particular this is true for items (usually not only just the material or the elements but also of information) that are generated by the previous step of the process and become input elements for the following ones.

Indeed it can be considered a particularly critical element the fact that the uncertainty associated with the information that is generated by a process is, in a certain sense, amplified when information affected by uncertainty serves as an input for the following process; this progressive amplification of uncertainty may generate in the process as a whole a level of uncertainty particularly important but at the same time difficult to predict and manage (Villemeur 1992).

In essence, if the input data of a process are highly uncertain it is plausible that, by taking decisions upon these information, in the following steps such uncertainty is further amplified. A single fault—in, for example, the definition of a unit price that must be associated with a quantity of machining particularly important—may result in a large error in the evaluation of the intervention as a whole (Cerutti and Paganin 2012).

The information that is generated or processed in different sub-processes are of different nature and complexity, and each may have a degree of uncertainty and variability that can influence, positively or negatively, the results of the construction process.

Considering this scenario from the point of view of management it can be said that in a series of moments choices are made that have major repercussions on the operation and maintenance phase, from the planning of the intervention, to the choice of architectural configuration, to the final design. The decisions taken in each of these moments go to influence, with a cascade effect, building behaviour during the operation and management phase and, if left unchecked, with no control on the final result.

In these crucial early phases of the process the actors that have the main influence and control over the process, are the client and designers, which therefore have also the major responsibility for the final building behaviour, while those in charge of running buildings' operation and carrying out the due maintenance have a sense of the quality of design in usage.

With reference to this scenario there is a significant interest in looking at the three phases that play the most prominent role in determining the satisfaction of the promoter's needs throughout the building process: the brief, the design and the operation and maintenance.

(a) The Role of the Brief Phase with Reference to the Satisfaction of the Promoter Needs Throughout the Building Process

The brief document is a document prepared by the head of the procedure. It contains technical and administrative in-depth specifications for the design and information such as the situation pre-intervention, the general objectives to be pursued and the strategies to achieve them, the needs to be satisfied, the rules and technical standards to be met, the specific constraints of the law linked to the specific context in which the intervention is provided, the functions that the intervention must fulfil, the technical requirements that must be met, the impact of the intervention on the environment, the phases of the design process to be developed and their logical sequence, as well as the relative times of execution, the levels of the design elaborated both graphic and descriptive to be prepared, the budget limits to be observed, the estimated costs and sources of financing to be used (Maestosi 2004).

In final analysis, the brief document is the input that instructs the design (which is the most prominent and crucial decisional action of the process). Therefore, any lack of clarity or error in preparing the brief is a source of uncertainty on the future phases of design, construction and use.

(b) The Role of Design with Reference to the Satisfaction of the Promoter's Needs Throughout the Building Process

The input information required for the design phase are: brief document and programming outcomes, while the output are: design drawings (at the different levels of detail) and contract framework.

The design process is highly exposed to risks. Project design risks evolve from misunderstanding project objectives and threats to achieving these objectives. This is particularly important for aspects relating to design brief, whole life objectives and parameters, design programme, whole life cycle cost (WLC) plan, time constraints and quality issues. The success of understanding and translating project whole life objectives into a reality depends on the design team's in-depth knowledge of the whole life-cycle process and associated sources of risk. This situation is further complicated by the fact that design processes are highly interactive and involve an extensive exchange of information between design stakeholders (Preiser and Vischer 2005).

In final analysis then, what is decided in the design is nothing less than what will ideally be realized and then used. It should be the translation in a physical output of the needs expressed by the clients in the brief, and it is required to satisfy that set of needs throughout the whole building lifecycle.

The ability of designers to envision effects of their decisions in the phase of operation and management, and to design according to that, determine the level of risks of design solutions over the satisfaction of clients' need in the long term. Design is without doubt the key decisional moment of the whole building process and the vast majority of problems that can occur in carrying out operation and due maintenance can be tracked back to design errors.

(c) The Role of the Operation and Maintenance Phase with Reference to the Satisfaction of the Promoter Needs Throughout the Building Process
Operation and management phase is one of the parts of the building process that concerns the greater length of time within a building life cycle.

Operation and maintenance is the phase in the future from the decisional moment of design when the inertial outcomes of early decision appear. For this reason, in this phase feedback can be gathered to reveal the quality of decisions taken upstream.

Programming tasks and expectations for the management and maintenance are considered the moments of connection between the construction phase and the operational phase of the building process.

The opportunity to consider the parameters of durability, cost of use, maintainability and substitutability, overall cost, in addition to traditional design, comes from the changed conditions in the construction market, and the move to more rational use of existing resources and optimization of investment throughout the construction life cycle (Molinari 1999).

The ease, time and costs of carrying out due maintenance largely depend on the adequacy of the maintainability of the building and building components. Therefore the phase of operation and maintenance can be considered the only moment in which the quality of design in use can be evaluated.

In conclusion, if designers properly consider the needs for the use and for maintenance, then the quality of buildings during the phase of operation and maintenance will be totally satisfactory, otherwise a number of problem will arise: from high cost of maintenance to discomfort in living. For this reason feedback gathered from the phase of operation and maintenance can serve to prove quality of design with reference to long-term objectives.

1.3 Learning by Using. The Role of Feedback Information to Improve the Effectiveness of Brief and Design in Managing the Risks Over the Long Term

As mentioned in the previous chapters, the operation and management phase is the time after the decisional steps from which consequences of decisions taken upstream appear. At this stage quality of buildings is visible and measurable and, therefore, feedback can be gained and reported to decision makers. This process can be defined as learning by using. Learning by using is particularly useful for permanent building developers that can take advantage of past experiences to do better in the future.

The ability of organizations to capitalize on feedback from the operation and management phase relies on whether those organizations are permanent rather temporary. Even if in constructions industries promoters are generally temporary multi-organization (Sinopoli 1997), examples can be found of permanent organizations that promote and manage several building initiatives. In particular in Italy significant example in this sense are:

- Banks and insurance companies, that own and manage several offices worldwide.
- Ministry of Army, which owns and manages several barracks, through the army inspectorate for infrastructure.
- Ministry of Health, which owns and manages many districts and hospitals to the local "ASL".
- The Italian Church, which owns and manages an important patrimony over the whole national territory.

With reference to organizations like these learning from the use in order to improve future processes of design and management are particularly valuable.

The search for a "fil rouge" between design, construction and problems in management would require a purely diagnostic and in-depth analysis, whose complexity is mainly due to the fact that it involves actions and phenomena that happen at very different times from each other. Where the causal link is between events and objects, which are physically and temporally separated, is always fragmented and confused.

A further difficulty lies in the necessity to reconstruct the overall "space configuration" in which failure events occur, where such a configuration may change substantially with changes in techno-typological characteristics of each building, its use and usage profiles, the environment in which it is settle.[2]

The cost of an analysis of this nature is often so high as to be completely unaffordable. It can be of great help then to use the historic "memory" of these events (Arrow 1962).

The effectiveness of such a knowledge-management tool on the one hand is obviously proportional to the amplitude of the time horizon in which this "memory" is progressively consolidated in integrated flows of information. On the other hand, it is conditioned by the ability to filter the data from the "memory" of the people towards a systematic framework of information that is organized and easy accessible (Rosenberg 1982). The framework of information can only be built through a systematic collection of information during the operation and management phase which, properly collected and organized, helps to form a base of knowledge useful to educate and inform briefs and designs of future projects (Talamo 2003).

---

[2] For example through the application of methods "Failure Mode and Effects Analysis" (FMEA), or "Failure Mode Effects and Criticality Analysis" (FMECA) and the subsequent construction of adequate "fault trees" (or "errors trees").

In the final analysis, the collection of data from the phase of operation and management serves to verify the quality of decisions taken upstream and to point to what could have been done better. This process of learning by using is useful for several figures within the building process, in particular for:

- Designers, in order to understand what set of choices causes most of the problems in use and in maintenance.
- Clients, in order to understand what important aspects haven't been both specified in the brief and verified in validation.

Parties that have had the possibility to see the impact of their mistakes and to track their origin are more likely than others to not repeat the same mistake again. With reference to this aspect it should be considered that designers are professionals of the building process and, therefore, they are likely to repeat the process many times. Clients instead can be of many different types but, as mentioned in this chapter, there are numerous continuous promoters that can fruitfully learn from past experience to do better in future.

References

K.J. Arrow, The economic implications of learning by doing. Rev. Econ. Stud. **29**, 155–173 (1962)

A. Cerutti, G. Paganin, *Risk management per l'edilizia* (Dario Flaccovio editore, Palermo, 2012)

G. Ciribini, Durabilità e problemi manutentivi nelle attività di recupero, in: Recuperare n.6, a.II, luglio-agosto (1983)

B. De Finetti, *L'invenzione della verità* (Raffaello Cortina editore, Milano, 2006)

B. De Finetti, *Theory of Probability. A Critical Introductory Treatment* (Wiley, Chichester, 1990)

R. De neufville, S. Scholtes, *Flexibility in Engineering Design* (MIT Press, Cambridge, 2011)

G. Gigerenzer, *Reckoning with Risk: Learning to Live with Uncertainty* (Penguin Books, London, 2002)

P.C. Maestosi, *La manutenzione nel processo edilizio* (Alinea editrice, Firenze, 2004)

M. Marseguerra, E. Zio, *Basic of Monte Carlo Method with application to System Reliability* (LiLo-Le Verlag, Hagen, 2002)

E. Mills, *Building Maintenance & Preservation. A Guide to Design and Management* (Butterworth Heinemann, Oxford, 1994)

C. Molinari, in *Il nuovo quadro di riferimento tecnico-normativo per la manutenzione*, ed. by Curcio Manutenzione dei patrimoni immobiliari. Modelli, strumenti e servizi innovativi (Maggioli, Rimini, 1999)

C. Molinari, *Procedimenti e metodi per la manutenzione edilizia* (Esselibri editore, Napoli, 2002)

G. Paganin, *L'acquisizione delle informazioni per la manutenzione dei patrimoni immobiliari* (Sistemi Editoriali, Napoli, 2005)

N. Rosenberg, *Inside the Black Box: Technology and Economics* (Cambridge University Press, New York, 1982)

N. Sinopoli, *La tecnologia invisibile. Il processo di produzione dell'architettura e le sue regie* (Franco Angeli, Milano, 1997)

C. Talamo, *Il sistema informativo immobiliare* (Sistemi editore, Napoli, 2003)

C. Talamo, *Procedimenti e metodi della manutenzione edilizia*, vol. 2 (Sistemi Editoriali, Napoli, 2010)

A. Villemeur, *Reliability, Availability, Maintainability and Safety Assessment*, vol. 1 (Wiley, Chichester, 1992)

E. Zio, *Notes from the Seminar of Mathematics Culture* (Politecnico di Milano, Milano, 2012)

Standards and Laws

International Organization for Standardization, 2005. UNI EN ISO 9000:2005 Quality Management Systems. Fundamentals and vocabulary

Chapter 2
Uncertainty, Risk and Risk Management

Abstract The aim of this chapter is to present the risk management process as a method to control risks over building use and maintenance, by estimating the uncertainty over the most important objectives. Risk management deals with the issue of uncertainty. In particular it aims to reduce uncertainty by envisioning possible scenarios and making forecasts on the basis of what it is considered probable within a range of possibilities. For this reason in order to properly adapt the risk management process to the field of architectural design it is necessary to understand the meaning and the relationship between: uncertainty, probability, range of possibilities and foresight. Moreover, it is also necessary to make clear the state of the art with reference to risk management process, as regards: terms, definitions, steps and methods. The risk management process has been defined many times and many different versions have been given over the last decades. In the current context the voluntary standards have unified terms, definitions, frameworks and steps of the process of risk management. The framework offered by the International Standards is particularly important in order to systematize knowledge from literature from various fields and to create a common ground to deal without either misunderstanding or ambiguity. Finally, applications of the risk management process in various context are reported and is introduced the possibility of using such approach in architectural design to control risks over building use and maintenance is introduced.

Keywords Risk management · Uncertainty · Probability · Risk · Risk assessment · Monte Carlo simulation

2.1 Certainty, Uncertainty and Risk

In almost all circumstances, and at all times, we all find ourselves in a state of uncertainty. Uncertainty in every sense. Uncertainty about actual situations, past and present (this may stem for either a lack of knowledge and information, or from the incompleteness or unreliability of the information at our disposal; it may also

C. Martani, *Risk Management in Architectural Design*,
PoliMI SpringerBriefs, DOI 10.1007/978-3-319-07449-8_2

stem from a failure of memory, either ours or someone else's, to provide a convincing recollection of these situations).

Uncertainty in foresight: this would not be eliminated or diminished even if we accepted, in its most absolute form, the principle of determinism. In fact the abovementioned insufficient knowledge of the initial situation and of the presumed laws would remain. Even if we assume that such insufficiency is eliminated, the practical impossibility of calculating would remain. Uncertainty in foresight performance of complex organisms interacting with surrounding environment, like buildings are, is obviously unavoidable.

Uncertainty in the face of decisions: more than ever in this case, compounded by the fact that decisions have to be based on knowledge of the actual situation, which is itself uncertain, to be guided by the prevision of uncontrollable events, and to aim for certain desirable effects of the decisions themselves, these also being uncertain.

The uncertainty is an issue that has been defined and addressed in various ways in the history and with approaches of various types (for example with the statistics and the calculation of probabilities). The notion of uncertainty is referred, by definition, to those events that are impossible or difficult to predict, and can be distinguished in two types: the *random uncertainty* (Zio 2012)—which depends on the characteristics of the system and can generally be traced back to random variations, fluctuations and occasional stochastic phenomena—and the *epistemic uncertainty* (Zio 2012)—that depends, instead, on defects of knowledge and perception of the analyst. Unlike aleatory uncertainty, the epistemic one can be not only estimated, but also treated and reduced (De Finetti 2006).

The expression "randomly" could be held to coincide with the theoretical unavailability of a conceptual model to configure the uncertainty considered related to the role of the observer who chooses the configuration in which to assess the uncertainty (Bruno et al. 2006). However, the uncertainty that we are interested in this work is that, at the bottom of the science itself, of the possibility to make inferences in such a way that an effective approach, a theory, are repeatable in different conditions while maintaining the same effectiveness to the change of the context and of the overall configuration (Minati 2009).

Within the building construction field uncertainty is traditionally considered and treated mainly in managing time and budget during construction, while is barely considered when it comes to performance. The only segment within building service studies that contemplate uncertainty is that of reliability studies (Harr 1987).

Either in foreseen or in facing decisions, the concept of uncertainty is strongly related with that of risk. Indeed risk, which have been given numerous definitions and classification (in 1997 a review of research interests in risk in the social and economic area in academic institutions in the United Kingdom identified 170 areas related to risk—conducted by the Department of Economics of the University of Newcastle (1997)—and almost all of them have an its own definition), is defined by the International Standard ISO GUIDE 73:2009 as the effect of uncertainty on objectives. Meaning by effect a deviation from an expected outcome, either positive or negative. Objectives can be referred to different aspects, such as financial, health and safety, and environmental goals, and can apply at different levels (such as strategic, organization-wide, project, product and process) but that doesn't change the meaning of risk.

The definition of risk given by the ISO standard strongly underlines how the lack of confidence over the likely of an event to happen—in relation to its eventual magnitude—define the risk of the event.

Risk is often characterized by reference to potential events and consequences, or to a combination of these. Indeed, risk is often expressed in terms of a combination of the consequences of an event (including changes in circumstances) and the associated likelihood of occurrence.

Even though for the ISO GUIDE 73 effect of risk can be either positive or negative, commonly risk is intended as the chance that harm will occur. With reference to that, risks can be distinguished for their impact among: direct, indirect, and consequential.

This distinction is based on whether the consequences of the event directly affect the assets, such as in case of fire or accident at work, or produce indirect effects to them, such as in case of unavailability of elements due to errors in maintenance, or consequences on reputation.

Moreover a distinction in different categories of risks can also be done. Below the main categories of risks proposed in literature are reported and analysed. Risks are classified in: pure and speculative, internal and external, strategic and operational, systematic and specific.

In such a complex and articulated process, as the building process is, all the types risk listed above are involved. And it is therefore crucial to control the uncertainty over the expected results of design proposals to manage the risk of promoters and users' objectives.

2.2 Risk Management Process and Current Contexts Where It Is Used to Reduce Uncertainty

Risk management process has been defined many times and many different versions have been given with a variable number of phases.

Among all version proposed in the past some prominent ones that can be cited are these of: Perry and Hayes (1985), Carter et al. (1994), Kliem and Ludin (1997) and Baker et al. (1998).

In the present chapter scope, principles and steps of the risk management process are assumed as they are presented in the International Standard ISO 31000:2009, which has collected all previous contribution and set a frame that has become a point of reference.

The risk management process, as intended in the International Standard ISO 31000, is defined as a set of coordinated activities to direct and control an organization with regard to risk and is made of four steps: communication and consultation, establishing the context, risk assessment, risk treatment, and monitoring and review. Figure 2.1 shows process with all its phases and relation between phases.

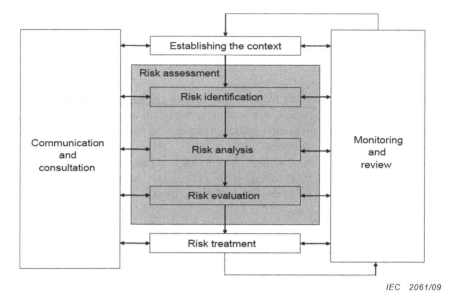

IEC 2061/09

Fig. 2.1 Risk management process in the ISO 31000

ISO 31000 suggests that organizations that want have an effective risk management process have to comply—at all levels—with the principles below:

(a) Risk management creates and protects value. Risk management contributes to
    the demonstrable achievement of objectives and improvement of performance
    in, for example, human health and safety, security, legal and regulatory
    compliance, public acceptance, environmental protection, product quality,
    project management, efficiency in operations, governance and reputation.
(b) Risk management is an integral part of all organizational processes. Risk
    management is not a stand-alone activity that is separate from the main
    activities and processes of the organization. Risk management is part of the
    responsibilities of management and an integral part of all organizational
    processes, including strategic planning and all project and change manage-
    ment processes.
(c) Risk management is part of decision making. Risk management helps decision
    makers make informed choices, prioritize actions and distinguish among
    alternative courses of action.
(d) Risk management explicitly addresses uncertainty. Risk management explic-
    itly takes account of uncertainty, the nature of that uncertainty, and how it can
    be addressed.
(e) Risk management is systematic, structured and timely. A systematic, timely
    and structured approach to risk management contributes to efficiency and to
    consistent, comparable and reliable results.

(f) Risk management is based on the best available information. The inputs to the process of managing risk are based on information sources such as historical data, experience, stakeholder feedback, observation, forecasts and expert judgment. However, decision makers should inform themselves of, and should take into account, any limitations of the data or modeling used or the possibility of divergence among experts.

(g) Risk management is tailored. Risk management is aligned with the organization's external and internal context and risk profile.

(h) Risk management takes human and cultural factors into account. Risk management recognizes the capabilities, perceptions and intentions of external and internal people that can facilitate or hinder achievement of the organization's objectives.

(i) Risk management is transparent and inclusive. Appropriate and timely involvement of stakeholders and, in particular, decision makers at all levels of the organization, ensures that risk management remains relevant and up-to-date. Involvement also allows stakeholders to be properly represented and to have their views taken into account in determining risk criteria.

(j) Risk management is dynamic, iterative and responsive to change.

(k) Risk management continually senses and responds to change. As external and internal events occur, context and knowledge change, monitoring and review of risks take place, new risks emerge, some change, and others disappear.

(l) Risk management facilitates continual improvement of the organization. Organizations should develop and implement strategies to improve their risk management maturity alongside all other aspects of their organization.

The risk management process (RPM) aims to control the uncertainty over important targets. Therefor it is a very broad approach that can be involved in any fields and any organizations that have important goals to achieve. Events that can be considered risky are of a huge variety as well. For example risks can be:

- *physical/material*—loss due to fire, corrosion, explosion, structural defect, war;
- *consequential*—loss of profits following fire, following theft;
- *social*—changes in public opinion, expectations of work force, greater awareness of moral issues (e.g. environment);
- *legal liabilities*—tortious liabilities, statutory liabilities, contractual liabilities;
- *political*—governmental intervention, sanctions, acts of foreign governments, inflationary/deflationary policies, export/import restrictions, trading alliances, changes in legislation;
- *financial*—inadequate inflation forecasts, incorrect marketing decisions, credit policies;
- *technical*—increased technology in manufacture, communications, data handling, interdependency of manufacturers, methods of storage, stock control and distribution.

Many fields exist where the risk management process is applied and these are also heterogeneous for goals and interests. Some significant areas where the use of

RMP is consolidates are, for example: army, finance, engineering (in particular: energy production, navy, aircraft and infrastructure) and the manufacturing industries.

Risk management process can potentially offer a strong contribution in reducing uncertainty (and so risks) in the current process of building design.

Nevertheless, given all principles listed in the ISO 31000 is clear that, as opposite to all fields were the RMP is already consolidate, the building process—which is characterized from a number of temporary multi-organization (Sinopoli 1997)—doesn't currently guarantee the conditions required transfer risk management techniques from other sector.

There is, therefore, much scope to create required conditions to include RMP in the key phases of the building process: the decisional moment of briefing and designing and the checking phase of operation and management.

2.3 Risk Assessment Techniques: Scope and Types

Within the context of the whole risk management process a fundamental role is played by the risk assessment phase (RA), in which a level of risk is defined as the degree of uncertainty over the most relevant objectives.

The RA phases provides decision-makers and responsible parties with a good understanding of risks that could affect achievement of objectives, and the adequacy and effectiveness of controls already in place. This provides a basis for decisions about the most appropriate approach to be used to treat the risks. The output of risk assessment is an input to the decision-making processes of the organization. The risk assessment is an overall process which includes 3 steps: risk identification, risk analysis and risk evaluation:

(a) Risk Identification
Risk identification is the process of finding, recognizing and recording risks. The purpose of risk identification is to identify what might happen or what situations might exist that might affect the achievement of the objectives of the system or organization. Once a risk is identified, the organization should identify any existing controls such as design features, people, processes and systems.

(b) Risk Analysis
Risk analysis is about developing an understanding of the risk. It provides an input to risk assessment and to decisions about whether risks need to be treated and about the most appropriate treatment strategies and methods.

Risk analysis consists of determining the consequences and their probabilities for identified risk events, taking into account the presence (or not) and the effectiveness of any existing controls. The consequences and their probabilities are then combined to determine a level of risk.

(c) Risk Evaluation

The purpose of risk evaluation is to assist in making decisions, based on the outcomes of risk analysis, about which risks need treatment and the priority for treatment implementation.

Risk evaluation involves comparing the level of risk found during the analysis process with risk criteria established when the context was considered. Based on this comparison, the need for treatment can be considered.

Examples of types of risk assessment methods available are listed in Fig. 2.2 where each method is rated to a level of applicability to risk identifications, risk analysis and risk treatment processes, according to its attributes.

With reference to the peculiar needs of long, multi-variable processes such as the building process is, among all techniques listed in Fig. 2.2 one is particularly useful to estimate the uncertainty over outcomes: the Monte Carlo method.

Monte Carlo (MC) method has the peculiarity of being a multi-factors technique, able to estimate the probability over the possible outcomes of a process dependent on many factors. In risk management over design choices the MC method is particularly useful to estimate the uncertainty over user requirements. The method avoids complicated computations of equations with many unknowns by simulating the actual situations to forecast based on the probability density function of all factors. Simulations are run by extraction of random numbers (Borgonovo et al. 2000).

2.4 Predictive Risk Management and Risk Management Through Process Monitoring and Responding. The Control of Uncertainty in the Built Environment

The aim of this chapter is to present possible approaches for managing performance risks in buildings over time. Two approaches are possible: predictive risk management, in which risks are envisioned and treated in advance, and risk management through monitor responding, in which risks are solved through real-time reacting to monitored conditions.

The risk management process can be divided into two main steps: a first step, during the decisional phases of the process, in which risks are identified, assessed and treated; and a second step, later on, in which the results are verified from information gathered from feedback.

In this framework almost all decisions to manage risks are taken in the first phase, while the collection of feedback serves to verify previsions, find mistakes and learn from errors. Following feedback, some corrections may be made to decisions taken in the first step, but in order to do so the feedback process must be fast enough to detect anomalies immediately, which is very rare.

Tools and techniques	Risk assessment process				
	Risk Identification	Risk analysis			Risk evaluation
		Consequence	Probability	Level of risk	
Brainstorming	SA[1]	NA[2]	NA	NA	NA
Structured or semi-structured interviews	SA	NA	NA	NA	NA
Delphi	SA	NA	NA	NA	NA
Check-lists	SA	NA	NA	NA	NA
Primary hazard analysis	SA	NA	NA	NA	NA
Hazard and operability studies (HAZOP)	SA	SA	A[3]	A	A
Hazard Analysis and Critical Control Points (HACCP)	SA	SA	NA	NA	SA
Environmental risk assessment	SA	SA	SA	SA	SA
Structure « What if? » (SWIFT)	SA	SA	SA	SA	SA
Scenario analysis	SA	SA	A	A	A
Business impact analysis	A	SA	A	A	A
Root cause analysis	NA	SA	SA	SA	SA
Failure mode effect analysis	SA	SA	SA	SA	SA
Fault tree analysis	A	NA	SA	A	A
Event tree analysis	A	SA	A	A	NA
Cause and consequence analysis	A	SA	SA	A	A
Cause-and-effect analysis	SA	SA	NA	NA	NA
Layer protection analysis (LOPA)	A	SA	A	A	NA
Decision tree	NA	SA	SA	A	A
Human reliability analysis	SA	SA	SA	SA	A
Bow tie analysis	NA	A	SA	SA	A
Reliability centred maintenance	SA	SA	SA	SA	SA
Sneak circuit analysis	A	NA	NA	NA	NA
Markov analysis	A	SA	NA	NA	NA
Monte Carlo simulation	NA	NA	NA	NA	SA
Bayesian statistics and Bayes Nets	NA	SA	NA	NA	SA
FN curves	A	SA	SA	A	SA
Risk indices	A	SA	SA	A	SA
Consequence/probability matrix	SA	SA	SA	SA	A
Cost/benefit analysis	A	SA	A	A	A
Multi-criteria decision analysis (MCDA)	A	SA	A	SA	A

Fig. 2.2 Tables of tools used for risk assessment and their applicability, were: *SA* stands for strongly applicable, *NA* for not applicable and *A* for applicable (IEC/ISO 31010:2009)

Risk management is a complex process characterised by a set of phases in which decisions such as planning and designing are taken early on, and the impact of those decisions analysed over a long period of use.

In line with these characteristics the risk management process seems to be particularly useful in controlling uncertainty over the long-term objectives for buildings and built environments. Indeed, the building process is characterized by a series of steps in which decisions that influence future quality are taken early on (brief and design), and where the quality of these decisions is revealed through time.

In addition to this it should be considered that in recent years the world of buildings and built environment construction and management has experienced a consistent increase in the use of systems for performance monitoring. Such systems have opened up the possibility of managing risks not only in a predictive manner, as in the traditional risk management process, but also in a reactive way, by using systems for sensing and responding.

In this sense two different methods seem to be possible for managing the risks in building and built environments: a predictive risk management and a risk management based on process monitoring and responding (Fig. 2.3). The two methods are not mutually exclusive: a project can benefit from both predictive risk management at first and then reactive risk management through its lifetime.

Risk Management Through Process Monitoring and Responding
One of the possibilities for reducing uncertainty in important objectives is through a system of monitoring and responding. Within this system, activities are constantly measured and actions are taken in real time to respond to any conditions that feedback indicates are not working properly. A system of process monitoring and responding requires two elements in order to be realized:

- Pervasive environmental sensor deployment, encompassing sensors, communications, massive data manipulation and analysis, data fusion with mathematical modelling, the production of outputs on a variety of scales from the local or domestic to the global, the provision of information as both hard data and user-sensitive visualization, together with appropriate delivery structures (Paulsen and Riegger 2006).
- The settlement of procedures for quickly responding to monitored conditions, consisting of a set of actions to be taken for each different scenario monitored.

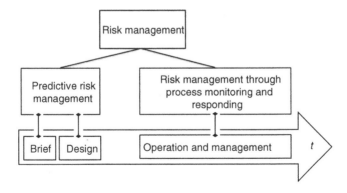

Fig. 2.3 Two possible risk management systems for buildings and built environments

Despite the fact that a risk management system based on process monitoring would eliminate the problem of uncertainty in previsions, such an approach has never been extensively used in building and building environments so far. The cost of pervasive sensor deployment has always been the obstacle to the use of this technique within the construction environment. But in recent years sensors have become ever cheaper and built environments have witnessed a diffusion of tools and systems for pervasive sensing that allow tests of real time sensing and responding from single building to urban level. And according to Resch et al. (2009) the decreasing cost of sensors will become particularly relevant in the United States if the environmental regulatory structure moves from a mathematical modelling base to a more pervasive monitoring structure.

At present, the use of sensors for risk management over long-term objectives mainly occurs at urban level, but recently it has also been applied to single buildings. Section 3.3 presents some cases of real time monitoring and responding at both urban and single building level. Examples are taken from my personal experience as research assistant at the Senseable City Laboratory of the Massachusetts Institute of Technology. In the project presented here, pervasive sensing was deployed in the built environment and systems for smart real-time responding were finalized.

References

S. Baker, D. Ponniah, S. Smith, Techniques for the analysis of risks in major projects. J. Oper. Res. Soc. (1998)

E. Borgonovo, M. Marseguerra, E. Zio, A Monte Carlo methodological approach to plant availability modeling with maintenance. Aging and obsolescence. Reliab. Eng. Syst. Saf. **67**, 59–71 (2000)

G. Bruno, G. Minati, A.G. Trotta, in *Uncertainty and the Role of the Observer*, ed. by G. Minati, E. Pessa, M. Abram, Systemics of Emergence: Applications and Development (Springer, New York, 2006), pp. 653–666

B. Carter, T. Hanock, J.M. Morin, N. Robins, *Introducing Riskman Methodology: The European Project Risk Management Methodology* (NCC Blackwell Ltd, Oxford, 1994)

B. De Finetti, *L'invenzione della verità* (Raffaello Cortina editore, Milano, 2006)

Department of Economics of the University of Newcastle (1997) *Risk and Human Behaviour*, ed. by J.C. Chicken, T. Posner (1998) The Philosophy of Risk (Thomas Telford, London)

M.E. Harr, *Reliability: Based Design in Civil Engineering* (McGraw-Hill Book Company, New York, 1987)

R.L. Kliem, I.S. Ludin, *Reducing Project Risk* (Gower Publishing Limited, Hampshire, 1997)

G. Minati, L'incertezza nella gestione della complessità. In Riflessioni sistemiche. Rivista italiana di studi sistemici, n°1 (2009)

H. Paulsen, U. Riegger, SensorGIS: Geodaten in Echtzeit. In: GIS-Business, vol. 8/2006, (Cologne, 2006) pp. 17–19

J.H. Perry, R.W. Hayes, Risk and its management in construction projects, in *Proceedings of the Institution of Civil Engineers*, Part I (1985), 78

B. Resch, M. Mittleboeck, S. Lipson, M. Welsh, J. Bers, R. Britter, C. Ratti, Urban Sensing Revisited: Common Scents—Towards Standardised Geo-sensor Networks for Public Health

Monitoring in the City (2009), http://senseable.mit.edu/papers/pdf/2009_resch_et_al__urban_sensing_CUPUM.pdf

N. Sinopoli, *La tecnologia invisibile. Il processo di produzione dell'architettura e le sue regie* (Franco Angeli, Milano, 1997)

E. Zio, *Notes from the Seminar of Mathematics Culture* (Politecnico di Milano, Milano, 2012)

Standards and Laws

International Organization for Standardization, 2009. ISO GUIDE 73:2009(E/F), Risk Management: Vocabulary

International Organization for Standardization, 2009. ISO GUIDE 73:2009(E/F), Risk Management: Vocabulary

International Organization for Standardization, 2009. IEC/ISO 31010, Risk Management: Risk Assessment Techniques

Chapter 3
Risk Management Through Process Monitoring: Reducing Uncertainty and Improving Risk Assessment Effectiveness Through Knowledge Gathering Over Time

Abstract The aim of this chapter is to propose the use of process monitoring on buildings to gain continuous feedback from the use, for both monitoring risks and learning by use. During this phase the actual performance of what has been predicted is measured in order to check the quality of forecasts. Eventual differences between performance expected and these measured would call either for a review on previsions, or for an intervention on the object measured. For monitoring of building performance data and the possibility for more data abounds but currently only few are the organizations prepared to actually gather the available information and use it for building management. Moreover monitoring techniques on building are often techniques to punctually check the level of performances and not actual continuous monitoring of them. To establish a system of building continuous monitoring that allows ready response, the systems of monitoring of processes can be adopted. Three examples of real-time monitoring and responding system, based on a monitoring of process, are reported from projects of the MIT | Senseable City Laboratory.

Keywords Building performance · Process monitoring · Rreal time monitoring · Monitoring and reviewing · Sensing and responding

3.1 Monitoring and Reviewing: Risk Assessment Updating Over Time

In both approaches to conduct risk management, (risk management through process monitoring and responding and predictive risk management), feedback-gathering from the operational phase is a key action. The first for real-time intervention and the second to better instruct future designs.

No risk management process can exist without a monitoring and reviewing phase, as indicated in the ISO 31000. Monitoring and reviewing is a set of activities that link risk management to other management processes and facilitate better risk

© The Author(s) 2015
C. Martani, *Risk Management in Architectural Design*,
PoliMI SpringerBriefs, DOI 10.1007/978-3-319-07449-8_3

management and continuous improvement. Continuous monitoring and reviewing of risk is an important part of implementation, particularly for large projects or those in dynamic environments. It ensures new risks are detected and managed, and that action plans are implemented and progress effectively. One very common type of risk monitoring and reviewing, particularly in building construction, consists of fulfilling periodically a risk watch list that contains the major risks that have been identified for the project or the process to be monitored. In complex project more than one responsible hold a different risk watch list (Cooper et al. 2004).

The international Standard ISO guide 73:2009 defines the terms monitoring and review as the continual checking, supervising, critically observing or determining status in order to identify change from the performance level required or expected.

In this sense, reviewing can be considered the activity undertaken to determine the suitability, adequacy, and effectiveness of the subject matter to achieve established objectives. And given its role within the risk management process, risks and controls should be monitored and reviewed on a regular basis to verify that (IEC/ISO 31010:2009):

- assumptions about risks remain valid;
- assumptions on which the risk assessment is based, including the external and internal context, remain valid;
- expected results are being achieved;
- results of risk assessment are in line with actual experience;
- risk assessment techniques are being properly applied;
- risk treatments are effective.

Risk monitoring and reviewing functions to continually improve the framework for managing the risks. To that end, in order to ensure that risk management is effective and continues to support organizational performance, the quality of reviewing would depend to the capability of an organization to conduct a series of activities:

- measure risk management performance against indicators, which are periodically reviewed for appropriateness;
- periodically measure progress against, and deviation from, the risk management plan;
- periodically review whether the risk management framework, policy and plan are still appropriate, given the organizations' external and internal context;
- report on risk, progress with the risk management plan and how well the risk management policy is being followed; and
- review the effectiveness of the risk management framework.
- As part of the risk management process, risks and controls should be monitored and reviewed on a regular basis to verify that:
- assumptions about risks remain valid;
- assumptions on which the risk assessment is based, including the external and internal context, remain valid;
- expected results are being achieved;

- results of risk assessment are in line with actual experience;
- risk assessment techniques are being properly applied;
- risk treatments are effective.

Monitoring and reviewing link risk management to other project management processes. Continuous monitoring and review of risks is an important part of implementation, particularly for large projects or those in dynamic environments. It ensures new risks are detected and managed, and that action plans are implemented and progressed effectively.

The project manager should maintain a risk watch list, containing a list of the major risks that have been identified for risk treatment action. For large projects, appropriate managers at each level of management in the project will maintain their own risk watch lists for their areas of responsibility.

Given all aims and conditions of monitoring and reviewing within the RMP, it is possible to argue that that final phase closes the circle of the process by connecting buildings' performance in use with promoters' and clients' needs.

3.2 Monitoring Techniques for Building Performances

As mentioned in Chaps. 1 and 2 the risk management process is not commonly adopted within the construction field at the present time and, by consequence, also the activities of monitoring and reviewing, as they are intended within the RMP, aren't much used during the building process.

Information from the operation and management phase are sometimes collected within the current practice of building construction (indeed, data and possibility for more data abounds in building), but not for a better understanding of the whole construction process. Indeed, there are few organizations actually prepared to gather information on buildings in use and, in the cases in which this happens, monitoring techniques on building are often techniques to periodically check the level of performances and not actual continuous monitoring of them.

The most common technique to monitor quality over time is condition monitoring (CM). This technique is largely used in industrial context and in recent years it has been transferred also to the building where, however, it is not as mature as in industry yet.

Condition monitoring is a key element in planned maintenance, allowing remedial action to be taken to avoid the consequences of high cost and lost time of component failure. On plants and equipments, CM is based on trend analysis and regular sampling.

Condition monitoring in construction can be considered as the set of procedures required to evaluate the current conditions of a system or apparatus, obtained by the use of techniques ranging from the use of complex computerized instrumentation, to the exploitation of human sensitivity, in order to prevent faults and to implement the maintenance only in the presence of a potential fault and when it is most convenient according to the production program (Williams et al. 1994).

It is, in other words, about carrying out comparative, periodic or continuous measurements of parameters that are believed to well represent the conditions of the component or system under analysis, and therefore, allows assessing the current situation and the future trend (considering a possible deterioration).

The concept of "condition monitoring" was first proposed in the 1970s, when it became clear that the usefulness of this continuous detection process and subsequent data analysis of the operational reliability, was to gather information on the overall population of components in service on the basis of which to undertake any changes to the maintenance plan or program (Fedele et al. 2004).

The available techniques for measuring quality in the environmental, functional and spatial fields are often different from those available in the technological field and from those needed to assess quality of procedures and processes.

In particular, in building construction it is difficult to measure quality over time because data gathering in the construction industry require, in most cases, that operators to have a good experience of the phenomena and failure modes on buildings. In other areas, in fact, data collection can be conducted on the basis of elementary observations on easily detectable aspects (such as eye color, genders, weight, high, etc.). On the other hand, in the case of building pathologies, the identification of phenomenon and their quantification, may not be immediate for people not involved in the construction sector and, in any case, interpretation of the phenomenon is unlikely to be unambiguous in the absence of a unique valuation criteria. For this reason it is necessary that operators participating in data gathering already have a basic knowledge of buildings. And even when the obstacle of the preparation of the observers is passed, other three fundamental difficulties have to be fixed for an effective and efficient use of condition monitoring on buildings (Morabito and Nesti 2000):

- The subjectivity of operations: A node of fundamental importance to be solved is to make the most unambiguous evaluations possible on pathological and nonpathological states of systems.
- The unification of criteria: For all failure modes under consideration it is necessary to establish evaluation criteria as homogeneous as possible in order to make unambiguous measurements.
- The volume of data: The construction industry is typically characterized by a lack of statistical data on the performances of materials. In addition, the collection of new data is also complicated because buildings do not ensure standardized conditions, as industry does, therefore it is hard to use data collected in one context to forecast for performances in other contexts.

Despite all these difficulties, data gathering on buildings it is not impossible. Indeed, Informative Systems for building management are examples of situations where building monitoring is commonly conducted.

This is the reason why the methods are distinguished depending on the mode of measurement adopted. In the text Technical Standards and Building Industrialization, five modes to measure quality are presented that differ from others in the procedure used to assign the value of quality. It follows a definition for each of the five type of measuring modes:

The first type includes those modes that use laboratory tests, or natural methods, understood as laboratory experimentation or testing on the behaviour of objects in place under given conditions: this is the case, for example, of the fire resistance tests on a given prototype.

Analogues are the methods of the second type, or semi-natural methods, which consist of evidences coming from the tests, like those of the first type, but in this case the conditions are altered during the tests in order to reduce the time and costs of the operation: this is the case, for example, for the test on the behaviour of components over time.

In the third type are included procedures that use mathematical models,[1] usually technical and experimental investigation of the physical behaviour: those, for example, for computing the heat losses or the static safety.

The fourth type includes methods[2] to verify the satisfaction of specific requirements: for example the presence of features that allow accessibility for disables.

Finally, in the fifth type are classified statistical methods that subjectively evaluate, in statistical terms,[3] the satisfaction of users (or a sample representation of users) to particular performance (Manfron and Zennaro 1988).

Despite all the mentioned methods for building monitoring there still is, in many cases, a significant concern related to reliability of tools for building monitoring. Indeed, not all available tools for performance and quality monitoring are characterized by the same degree of maturity, as demonstrated in a research on building quality monitoring conducted from the Institute of Architecture | University of Venice. In that study, a qualitative assessment was conducted on a sample of public housing, built in the Veneto Region (in accordance with law 457/78—also known as the Ten-Year Plan for the house).

The results of this study showed a clear difference of maturity of the various monitoring techniques. Indeed results suggested that environmental requirements relating to usability, flexibility, integration and helio-thermic comfort were wrongly assessed, raising some concerns on the effectiveness of the monitoring of some requirements related to physical and acoustic comfort (Manfron and Zennaro 1988).

Despite the difficulties in measuring performances, some cases where building condition monitoring techniques have been proposed and tested have occurred in the last decade: In the late 1990s in South Africa, subsequent to the National Health Facilities Audit of hospitals, condition assessments have evolved into a technology

---

[1] Mathematical methods: mathematical models are, in general theoretical models that simulate the performance object for verification. For example models to test behavior under load of structures.

[2] Verification methods: are methods to verify the presence of provisions in respect to precise requirements.

[3] Subjective methods for test statistics: they are "investigations Doxa" verification methods. Evaluate, with appropriate techniques, the degree of satisfaction of a sample of users with regard to determining performance of the building product.

that added the time dimension to strategic management and maintenance of buildings and related infrastructure (Pedro et al. 2008). A method for building conditions assessment within a common-agreed European checklist for building objects (sub-divided into types) were proposed in 2002 as part of a European project—the TOBUS project—to develop a tool for selecting Office Building Upgrading Solutions (Brandt and Rasmussen 2002).

In the construction field, the most common techniques of not-invasive monitoring of technological performance, in addition to visual inspection, are the following (Brunetti 2007): thermographic, endoscopic; magnetometer, sonic auscultation, ultrasonic, test with horizontal cylinder, single flat-ram, double flat-ram, tension measurement of metal chains, sclerometric, resistographic and pull-out.

In light to all this fragmentation of scopes and means for monitoring building performances, there seems to be scope for a much more solid monitoring and reviewing approach to serve for checking and improving a risk management process through the whole life of buildings.

3.3 Real-Time Monitoring and Responding

Real-time monitoring and responding is a system of monitoring that doesn't focus on the monitoring of elements every once in a while, but on monitoring of processes. It is, indeed, thought to continuously check the quality of performance not by actually checking the elements themselves, but by checking the performance of the system in use in order to quickly detect problems and act on them.

In Chap. 1 introduced how building process deals with uncertainty and how tools and methods for risk management act in order to control such factor.

Even though risk management process is based on a predictive approach in order to control hazards and avoid undesired events, the monitoring and review phases is embodied in the process.

Within the risk management process, monitoring and review is commonly seen as the set of activities that serve to verify previsions, find mistakes and learn from errors. But monitoring can also have another role; it can serve to act on ongoing processes, by responding to observed conditions.

In order to use feedback information to act on ongoing processes, monitoring must be in real time. It has to be, in other words, a system of sensing and reacting. This is the case, for example, of pervasive sensing monitoring linked to standardized procedures of reaction.

To the best of the author's knowledge, cases where real-time monitoring and responding (RTM&R) system are applied to building operation and management aren't available yet. The closest systems to RTM&R existing in building conduction at present are the Building/House Energy Management System (BEMS/HEMS) that coordinate energy request and energy capability in order to best fit needs and resources. BEMS are generally applied to the control of active systems, i.e. heating, ventilation, and air-conditioning (HVAC) systems, while also determining their operating times (Doukas et al. 2007), through advanced control techniques based on real time sensing and artificial intelligence (neural networks, fuzzy logic, genetic algorithms, etc.) (Kolokotsa et al. 2005; Kalogirou 2006). Though, as the cost of sensors doesn't allow a pervasive deployment yet, as would be required for an actual RTM&R systems, these technologies for providing more sustainable and efficient systems for building services management are not a reality yet.

Instead, cases where RTM&R systems are proposed can be found in other contexts, such as energy and safety management of large and complex real estates, as well as in traffic management. There is scope to assume that an understanding of the application proposed in these fields can potentially be beneficial to the construction industry for a knowledge transfer.

Three examples of real-time monitoring—through pervasive environmental sensing—and responding system are reported from projects I worked on in 2011 as a Research Assistant at the Senseable City Laboratory of the Massachusetts Institute of Technology.

The last of the three projects is conducted at a urban scale, while the first two refer to the buildings scale (large buildings, like university campus or a nuclear power plant, but still building scale).

The common ground of all these projects is the use of process monitoring, rather than condition monitoring on elements or components, to understand the state of systems. Process monitoring consists of the observation of operating characteristics of an item or group of items and it is normally used from operators, regulators and safeguards to understand if items are running efficiently and safely.

While the common practice of building monitoring is characterized by the use of conditions monitoring techniques, where items or group of items are checked periodically, process monitoring is much more efficient because it reveals the conditions without doing any inspection.

Indeed in all three projects from MIT Senseable City laboratory the running efficiency of systems—whether this means: state of safety on plants, occupancy rate and energy use in buildings, or use of parking spots in cities—is monitored, or even sensed in real time and without any punctual activity on items, but observing systems operating.

This approach is particularly interesting for the process of risk management on building performance. Indeed, the same principle applied to running buildings would lead to a much faster understanding of problems. For example monitoring the time and costs of maintenance interventions on buildings can serve as an indicator of the state of the element.

A summary of the three projects follow:

(1) Future ENEL

Future ENEL is a multi-disciplinary initiative at the MIT Senseable City laboratory aimed at researching futuristic sensor technologies applied to the safe construction and operation life cycle management of energy plants. Leveraging real-world case studies provided by leading Italian energy company ENEL, Future ENEL makes focus on development of technologies and techniques for the real-time sensing, modelling and advanced visualization of stationary and moving objects within power plants and related construction sites.

To this purpose of the project three element where realized: a real-time sensing system, a virtual control room and a nervous system.[4]

The system proposes is based on a sensing and reacting approach where information are gathered in real-time processed in the virtual control room so that feedbacks can be then diffused back on site.

(2) ENERNET

Enernet is a project in which a new method is proposed to measure activity, using WiFi connections as a proxy for human occupancy. Data on the number of WiFi connections and energy consumption (electricity, steam and chilled water) were compared for two buildings within the Massachusetts Institute of Technology's campus. The results of the study demonstrate: the operation of the heating, ventilation and air conditioning (HVAC) systems adhered more closely to factors other than occupancy i.e. external temperature, whilst a small part of the electricity levels did correlate with the occupancy. In order to present possible solutions to address the disconnect between the HVAC system and occupancy levels: using the spatial configuration of rooms to inform the heating/cooling supplied, alongside regulating the degree of control that individuals are able to exercise over the thermostat. These are just two examples from a suite of possible strategies that could be employed to not only improve the operation of the overall system, but also encourage behavioural change across campus. For example, if the range of a thermostat becomes increasingly limited with fewer occupants, individuals may be more inclined to share a space in order to achieve thermal comfort or they may compromise temporary discomfort in order to use the aforementioned space. In many ways providing this option to users brings to the fore the question of energy efficiency but also actively engages individuals by demonstrating practical solutions (Martani et al. 2012). In any case the thermal regulation react responding to real-time data on the occupancy rate.

(3) INFRASTRUCTURELESS PARKING SYSTEM

Infrastructureless Parking System (IPS) it's a project that aims to inform drivers of available parking without requiring the installation of sensors or smart meters on the ground; a system that is akin with Copenhagen's Green Growth plan and serves to reduce urban congestion while engaging citizens in maintaining an important public asset. A parking system delivering real time information and monitoring of the parking network in its entirety. The Infrastructureless Parking System improves the use of existing resources whilst also encouraging a change in the behavior of the systems users.

The Infrastructureless Parking System incorporates an inventory of all existing parking spaces in the City of Copenhagen, including their location, size, and degree of demand. It allows individual users to navigate through the urban environment to efficiently find parking using an augmented reality application. Once a space has

---

[4] Source: http://senseable.mit.edu/enel/.

been identified, individuals can electronically validate their parking through an online payment system. This in turn feeds back information to the system, removing the parking spot that has just been occupied from the data base of available parking spots for the time duration requested by the driver (Martani 2011). Also in this case the approach is based on a system of sensing and responding.

References

E. Brandt, M.H. Rasmussen, Assessment of building conditions. Energy Build. **34**(2), 121–125 (2002)

G. Brunetti, *Lezioni di tecniche di indagine non distruttive e monitoraggio* (Politecnico di Torino, Torino, 2007)

D.F. Cooper, S. Grey, G. Raymond, P. Walker, *Project Risk Management Guidelines: Managing Risk in Large Projects and Complex Procurements* (Wiley, Chichester, 2004)

H. Doukas, K.D. Patlitzianas, K. Iatropoulos, J. Psarras, Intelligent building energy management system using rule sets. Build. Environ. **42**, 3562–3569 (2007)

L. Fedele, L. Furlanetto, D. Saccardi, *Progettare e gestire la manutenzione* (McGraw-Hill, Milano, 2004)

S. Kalogirou, Artificial neural networks in energy applications in buildings. Int. J. Low Carbon Technol. **1**(3), 201–216 (2006)

D. Kolokotsa, K. Niachou, V. Geros, K. Kalaitzakis, G. Stavrakakis, M. Santamouris, Implementation of an integrated indoor environment and energy management system. Energy Build. **37**(1), 93–99 (2005)

V. Manfron, P. Zennaro, *Valutazione della qualità dell'edilizia residenziale pubblica nell'era veneta, in Qualità Norma e progetto* (Arsenale, Venezia, 1988)

C. Martani, *C9/Infrastructureless Parking System in Senseable City Guide to Copenhagen 2* (DUSP, MIT. SA + P press, Cambridge, 2011), pp. 51–58

C. Martani, D. Lee, P. Robinson, R. Britter, C. Ratti, ENERNET: Studying the dynamic relationship between building occupancy and energy consumption. Energy Build. **47**, 584–591 (2012)

G. Morabito, A. Nesti, *Valutare l'affidabilità in edilizia* (Gangemi editore, Roma, 2000)

J.A. Pedro, J.A. Paiva, A.J. Vilhena, Portuguese method for building condition assessment. Struct. Surv. **26**(4), 322–335 (2008)

J.H. Williams, A. Davies, P.R. Drake, *Condition Based maintenance and Machine Diagnostic* (Chapman & Hall, London, 1994)

Standards and laws

International Organization for Standardization, 2009. ISO 31000:2009, Risk management—Principles and guidelines

International Organization for Standardization, 2009. IEC/ISO 31010, Risk management—Risk assessment techniques

Websites and documents available online

MIT | Senseable City Laboratory. Future ENEL project [Online]. Available at: http://senseable.mit.edu/enel/

Part II
Methodological Experimentation: Proposal of a Dashboard for Risk Management in Design Through a Predictive Approach

Chapter 4
Risks Over Objectiveness in Building Process

Abstract The aim of this chapter is to propose a set of tools and methods to manage the risks for a number of objectives along the building process. All technological elements or sets of technological element can be traced to different requirements, as well as to a list of possible interventions. Risks for all requirements, both technological and user requirements, of all elements, strongly rely on two aspects: the amount of intervention on which the elements involved depend on and their degree of maintainability. For this reason, in order to manage the risks over long-term objectives in architectural design a set of tools and methods have been created. The final outcome of the process is a dashboard in which a level of risk is reported for all requirements. In order to define a level of risk for all requirements a set of tools and methods are introduced with the aim of evaluating both the importance and the uncertainty over all requirements, by correlating the need for maintenance and the maintainability of all elements involved with each requirement.

Keywords Risk over objectives · Maintenance · Maintainability · Performance over time · Technical requirements · User requirements · Procedure for risk management · Databases of interventions

4.1 Risks Over Objectives in Building Process: From Brief and Design to Operation and Maintenance

In accordance with the definition of risk given by the international Standard ISO 31000, as the combination of uncertainty and severity of consequences, the most risky and important objectives for the promoter of a building initiative are those whose satisfaction over time cannot be determined with certainty a priori.

Among all client objectives, those that by their nature are typically both important and uncertain and dependent on the behaviour of the buildings over a long time horizon are the objectives related to the operation and management phase.

Within the building process the performances required to satisfy over time the objectives of clients and users with reference to the operation and management

© The Author(s) 2015   37
C. Martani, *Risk Management in Architectural Design*,
PoliMI SpringerBriefs, DOI 10.1007/978-3-319-07449-8_4

phase are largely dependent on the maintenance activities, and the more the conditions required for conducting the maintenance activities are guaranteed, the more
likely it is that the maintenance will be carried out. In other words, the adequacy of
the maintainability of elements affects significantly the probability that their
requirements are kept satisfied over time. Thus, the adequacy of the maintainability
of elements, compared to the maintainability needed from their interventions, can
be taken as an indicator of the uncertainty around the satisfaction over time of the
requirements of those elements.

As introduced in Chap. 1, requirements that express the needs of clients with
reference to the operation and management phase, can be distinguished into two
types: technological requirements and user requirements. The technological
requirements are those related to specific technical elements, while user requirements are related to a space, a room or the entire building, and depend on the
performances of several elements.

Each technical element can be traced to different technological requirements
(Fig. 4.1), as well as to a list of possible interventions.

Moreover each intervention is characterized by a frequency and by particular
needs for maintainability. Indeed all interventions require specific conditions to be
guaranteed, ranging from needs for operations, to these for using tools and vehicles,

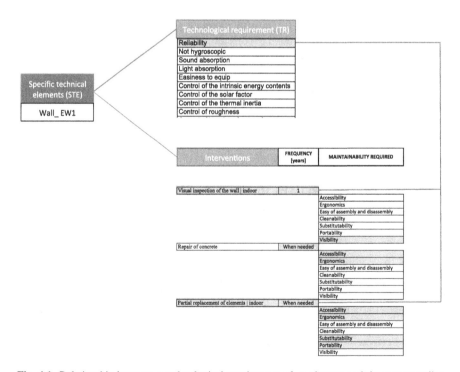

Fig. 4.1 Relationship between a technological requirement of an element and the corresponding
interventions

and the sum of all the requirements associated with all interventions of an elements compose the total amount of technical requirements of that element.

It follows that the degree of uncertainty of each technological requirement of an element strongly relies on two aspects: the amount of intervention it depends on (number and frequency) and the degree of maintainability of the pertinent element.

It can be assumed therefore, that the more the technological requirements of an element can be considered uncertain over time, the more the action needed to maintain the element is not guaranteed by the design.

An example of a map of connections is presented in Fig. 4.2 in which are linked: needs, requirements, technical elements (with associated level of maintainability) and interventions (with related frequencies and level of maintainability).

With reference to more complex requirements, such as the user requirements, which depend on the behaviour of various technical elements (Fig. 4.3), the satisfaction depends on the performances of several elements. Therefore in the same way in which the uncertainty around the maintenance of a technological requirement is strongly influenced by both the need for maintenance and the degree of maintainability of the pertinent element, the uncertainty around the maintenance of a user requirement is strongly influenced by these conditions for all the elements involved.

Requirements, however, either user or technological ones, do not all have the same importance. The importance of a requirement depends on a complex set of factors that can be quite effectively summarized in two elements: the presence of legal obligation (that includes many important cases, such as: the protection of health and safety of people, protection against environmental damage, etc.) and its degree of relevance with respect to the apparatus of needs, wishes and constraints expressed by the client and users.

In this sense, every requirement is extremely important if both mandatory by law and a priority for the promoter, and it is insignificant if neither the promoter consider it important nor it is mandatory by law.

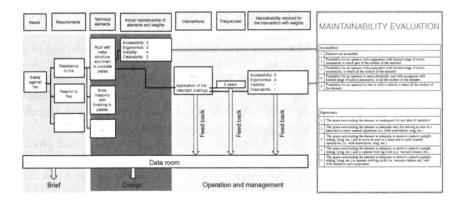

Fig. 4.2 Map of connections between needs, requirements, technical elements (with associated level of maintainability) and interventions (with related frequencies and level of maintainability)

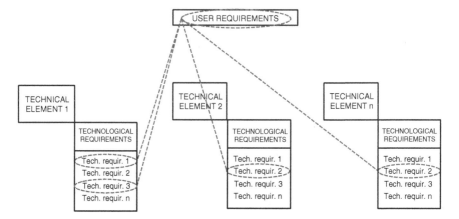

Fig. 4.3 Relationship between a user requirement and all technological requirements of the elements involved

It follows that the most risky requirements are those that, on one hand, derive from laws or are considered essential by the client and, on the other, are related to one or more elements whose maintainability is inadequate.

The phases within the building process that have a crucial role in determining the risk level of one or more requirements are the design and the brief phases. Indeed the design is the main decisional moment in the building process. In this context a number of choices are taken that have a decisive influence in determining both the need for maintenance and the level maintainability of the whole building and its parts and components: materials, techniques and types construction, as well as spaces dimension and shapes (just to name a few).

While the brief is the phase in which clients define goals and give technical and administrative in-depth specifications for the design, among which are: the needs to be satisfied, the rules and technical standards to be met, the specific constraints of the law linked to the specific context in which the intervention is provided, the functions that the intervention must fulfil, the technical requirements that must be met.

The performances of buildings over time, which determines the satisfaction of a framework of requirements, largely rely on the choices taken within these two decision steps made at the beginning of the building process. Instead, during the phase of operation and management the outcomes of the choices taken upstream (the indications given in the brief and their translations in spaces and objects conducted in design) appear and their effectiveness becomes visible.

The performance of buildings during the operations and maintenance phase provides useful feedback both to fix problems in running buildings and to enhance the preparation of future briefs and designs in order not to repeat the same mistake again. Feedback information is typically of two types: records of interventions performed with relative frequency, cost, duration, type of market, means and freight employees, and performance evaluations (where possible and appropriate).

4.2 Proposal of Dashboard Based on Tools and Procedures to Identify, Assess and Evaluate the Risks Over Objectives

In the present work the use of a dashboard of risks is proposed to assess and manage the level of risk over time of a set of requirements. The dashboard consists of two matrices: the matrix of risks of technological requirements (Fig. 4.4) and the matrix of risks of user requirements (Fig. 4.5), which are developed on the basis of a building breakdown structure at the level of technical elements, a database of technological requirements and a database of user requirements.

Building breakdown structure

The building breakdown structure was created by assuming the classes of technological units, the technological units and the classes of technical elements from the Italian Standard UNI 8290-1. Then a level of typological configuration of technical elements has been added that includes the more common technical elements. The list of typological configuration of technical elements was created from literature: (Kind Barkauskas et al. 1998; Natterer and Herzog 1998; Schunck et al. 1998; Schittich et al. 1998; Zaffagnini 1994).

Database of technological requirements

The database of technological requirements was derived from the requirements listed in the Italian Standard UNI 8290-2.

Database of user requirements

The database of user requirements was created by integrating multiple sources: the Italian Standard UNI 8289, that reported the classes of needs, the International Standard ISO 6241 and the International Existing building Code of ICC (2003).

The level of risk associated to the requirements of both types within the dashboard of risk ranges from 0—no risk at all—to 25—maximum risk—and it is computed by multiplying the degree of importance and the uncertainty of each requirement (Fig. 4.6).

In line with the mentioned procedure to compute risks, in order to create the dashboard other tools were prepared first: a tool to assign a degree of importance to requirements and an apparatus of tools and procedure to estimate the level of uncertainty over requirements.

Tool to assign a degree of importance to requirements

The tool that has been prepared to assign a degree of importance to requirements is based on the two matrices of the dashboard: the matrix for technological requirements (Fig. 4.7) and the matrix for user requirements (Fig. 4.8). In each matrix a degree of importance is given to all requirements according to both the promoter needs and to the legal obligations. The degree of importance can range from 1 to 5 where, 1 corresponds to irrelevant requirements, and 5 to fundamental requirements.

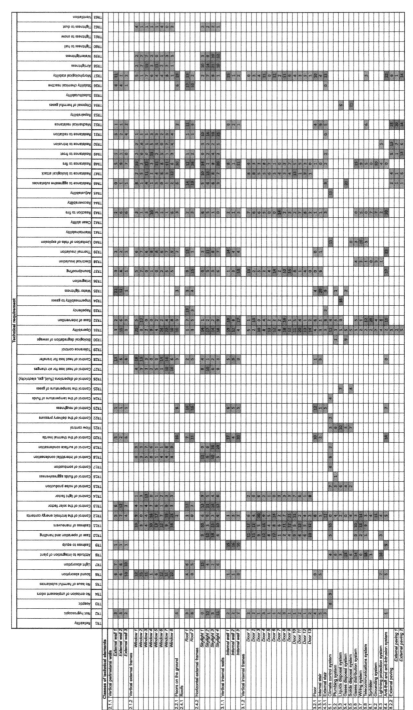

Fig. 4.4 Matrix of risks of technological requirements

	User requirements																							
	Morphological efficiency in relation to actions (both static and dynamic)	Evacuation in case of emergency	Control of risk of fire	Safety against pollutants	Protection against falling	Security against intrusion	Control of internal air temperature	Summer comfort	Control of orientation	Control of noise	Control of natural lighting	Control of ventilation	Control of odors	Resistance to external agents	Resistance to the stresses of exercise	Accessibility to people and things	Accessibility for people with disabilities	Furnishability	Practicability	Adaptability overtime (possibility of merging, splitting, spaces)	Functional integration	Dimensional integration	Control of pollution	Energy efficiency and use of natural energy
Building and building areas	UR1.1	UR2.1	UR2.2	UR3.1	UR4.1	UR5.1	UR6.1	UR6.2	UR6.3	UR7.1	UR8.1	UR9.1	UR9.2	UR10.1	UR10.2	UR11.1	UR11.2	UR12.1	UR12.2	UR13.1	UR14.1	UR14.2	UR15.1	UR16.1
The whole building	17	10	14	8	10	14	7	9	10	14	7	12	6	7	10	11	7	6	9	9	2	2	9	9
Area 1	18	10	15	3	10	15	4	4	4	25	6	14	4	8	4	5	4	4	15	1	1	1	8	8
Area 2	19	11	12	15	8	12	13	12	15	6	6	10	8	6	15	15	10	9	4	25	5	2	10	10

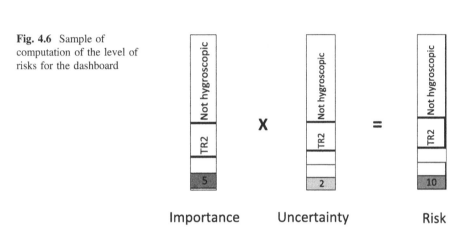

Fig. 4.5 Matrix of risks of user requirements

Fig. 4.6 Sample of computation of the level of risks for the dashboard

Tool to estimate a degree of uncertainty over requirements

In line with the previous tools, the tools to estimate a degree of uncertainty over requirements is composed of a matrix for technological requirements (Fig. 4.9) and a matrix for user requirements (Fig. 4.10).

Within this tool a degree of uncertainty is given to each requirement in a range from 1 to 5 (where 1 correspond to very low uncertainty, and 5 to very high uncertainty). As the uncertainty around the satisfaction of a requirement in the long term is largely influenced from the adequacy of the maintainability of pertinent elements, in order to estimate the degree of uncertainty around requirements—both technological and user requirements—a map of correlation was created between required and actual maintainability of elements. The less the maintainability meets the requested level, the more the satisfaction over time of pertinent requirements is evaluated to be uncertain.

In order to evaluate the degree of uncertainty on requirements due to the adequacy of maintainability a set of preparatory tools was created first:

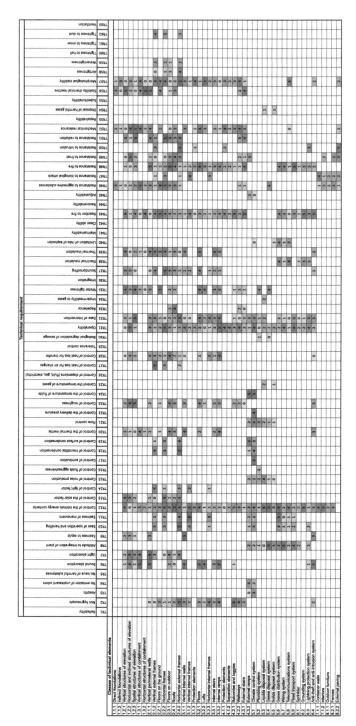

Fig. 4.7 Matrix of the degree of importance of technical requirements

	User requirements																							
Building and building areas	URL1.1 Morphological efficiency in relation to actions (both static and dynamic)	UR2.1 Evacuation incase of emergency	UR2.2 Control of risk of fire	UR3.1 Safety against pollutants	UR4.1 Protection against falling	UR5.1 Security against intrusion	UR6.1 Control of internal air temperature	UR6.2 Summer comfort	UR6.3 Control of orientation	UR7.1 Control of noise	UR8.1 Control of natural lighting	UR9.1 Control of ventilation	UR9.2 Control of odors	UR10.1 Resistance to external agents	UR10.2 Resistance to the stresses of exercise	UR11.1 Accessibility to people and things	UR11.2 Accessibility for people with disabilities	UR12.1 Furnishability	UR12.2 Practicability	UR13.1 Adaptability over time (possibility of merging, splitting, spaces)	UR14.1 Functional integration	UR14.2 Dimensional integration	UR15.1 Control of pollution	UR16.1 Energy efficiency and use of natural energy
The whole building	5	5	5	4	4	4	3	3	4	4	4	3	3	2	3	5	4	2	3	3	2	1	2	2
Area 1	5	3	5	3	5	4	2	2	4	5	4	3	3	2	2	5	4	2	3	1	1	1	2	2
Area 2	5	5	5	5	3	4	4	4	4	3	3	3	3	2	4	5	4	2	3	5	3	1	2	2

Fig. 4.8 Matrix of the degree of importance of user requirements

- A tool to evaluate maintainability of elements;
- A set of spreadsheets, to identify the requests of maintainability for all technological requirements of each given typological configuration of technical elements.

4.3 Tool to Evaluate the Maintainability of Elements

A tool to evaluate the maintainability of elements, on the basis of the features of the object and of their surrounding spaces, was created for 7 different classes of technical elements:

- A tool to evaluate the maintainability of external walls;
- A tool to evaluate the maintainability of internal walls;
- A tool to evaluate the maintainability of roofs;
- A tool to evaluate the maintainability of external vertical frame;
- A tool to evaluate the maintainability of internal vertical frame;
- A tool to evaluate the maintainability of floors;
- A tool to evaluate the maintainability of plants.

Each of the tools created aims to evaluate an object with reference to 7 factors of maintainability and is composed of 9 sheets (Molinari 1994; Molinari 1999; Molinari 2002; Zaffagnini 1994; Ciribini 1983; Curcio 1995; Curcio and D'Alessandro 1994):

(a) ID;
(b) Sheet to evaluate accessibility;
(c) Sheet to evaluate Ergonomics;
(d) Sheet to evaluate Ease of assembly and disassembly;
(e) Sheet to evaluate Cleanability;
(f) Sheet to evaluate Substitutability;
(g) Sheet to evaluate Portability (of the LRU—Lowest Replaceable Unit);
(h) Sheet to evaluate Visibility;
(i) Final summary.

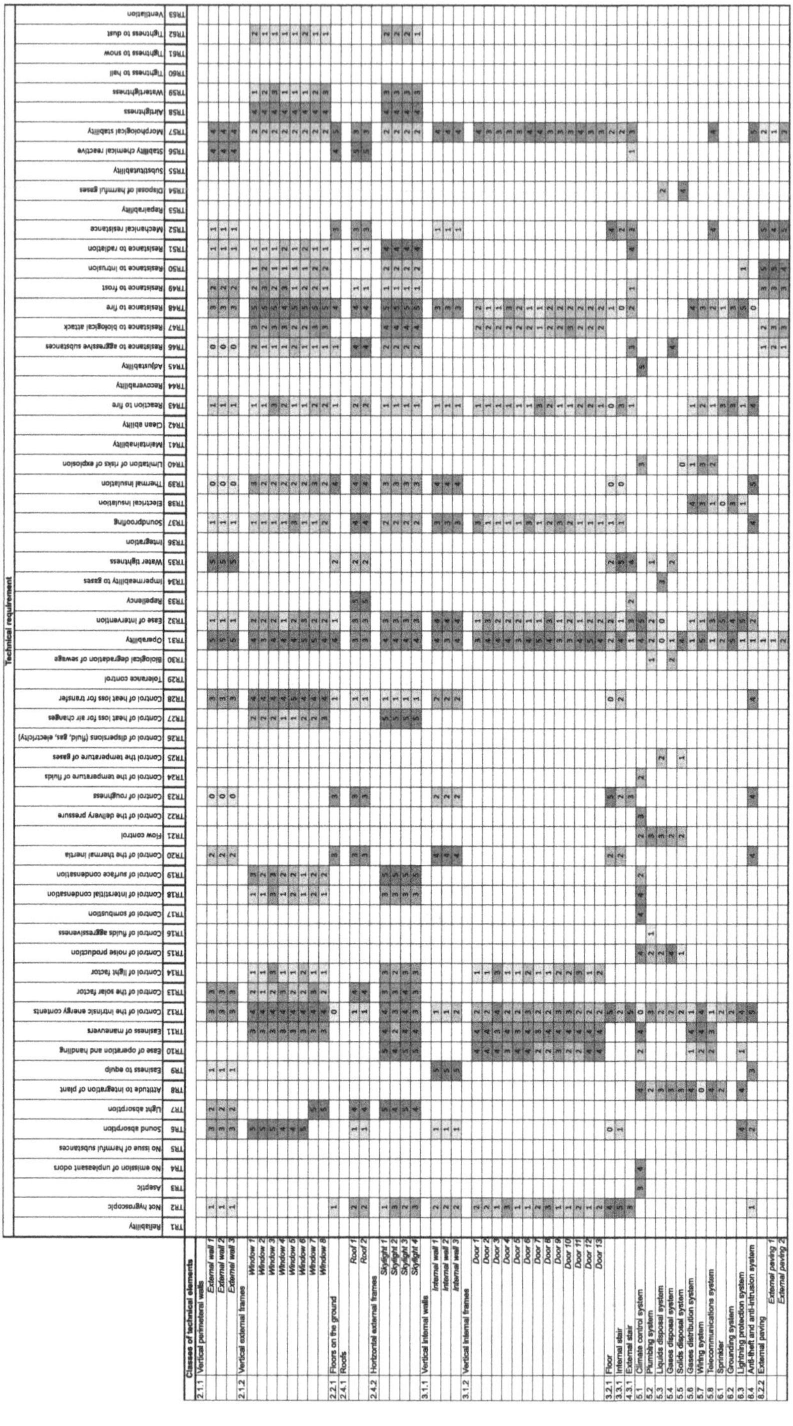

Fig. 4.9 Matrix of the degree of uncertainty of technological requirements

Building and building areas	User requirements																							
	UR1.1 Morphological efficiency in relation to actions (both static and dynamic)	UR2.1 Evacuation incase of emergency	UR2.2 Control of risk of fire	UR3.1 Safety against pollutants	UR4.1 Protection against failing	UR5.1 Security against intrusion	UR6.1 Control of internal air temperature	UR6.2 Summer comfort	UR6.3 Control of orientation	UR7.1 Controld noise	UR8.1 Control of natural lighting	UR9.1 Control of ventilation	UR9.2 Control of odors	UR10.1 Resistance to external agents	UR10.2 Resistance to the stresses of exercise	UR11.1 Accessibility to people and things	UR11.2 Accessibility for people with disabilities	UR12.1 Furnishability	UR12.2 Practicability	UR13.1 Adaptability over time (possibility of merging, splitting, spaces)	UR14.1 Functional integration	UR14.2 Dimensional integration	UR15.1 Control of pollution	UR16.1 Energy efficiency and use of natural energy
The whole building	3	2	2	2	2	2	3	2	3	2	4	2	3	3	2	2	3	3	3	1	2	4	4	4
Area 1	2	2	3	1	2	4	2	2	1	5	1	4	2	1	1	2	5	1	1	1	1	4	4	4
Area 2	3	2	2	3	3	3	4	4	2	3	3	3	5	4	2	4	1	5	2	2	2	5	5	5

Fig. 4.10 Matrix of the degree of uncertainty of technological requirements

Two sheets are reported as examples: the sheet to evaluate accessibility and the final summary:

Sheet b: accessibility
In Fig. 4.11 a sheet to evaluate the accessibility of an element is reported. The page is divided into three sections:

(a) A first section, named physical specifications of spaces and elements, in which all information from the last section of the ID page is automatically reported (location, surface, height, construction technique, breakdown structure, etc.)

(b) A second section, named specifications of spaces and tools for maintenances in which are listed: the size and characteristics of space for maintenance, the size and characteristics of tools and vehicles for maintenance

(c) A third section to evaluate the condition of the accessibility of the element. In this area both internal and external accessibility of the element are evaluated by assigning a level within a given range.

The internal accessibility can range from 0 to 2, where each value has the following meaning:

0. Indoor side of the element inaccessible;
1. Possibility for an operator to access to the internal space of the element;
2. Possibility for an operator to access to the internal space of the element with tools and parts and sub-components to substitute.

The external accessibility can range from 0 to 7, where:

0. Element inaccessible;
1. Possibility of accessing to the element only through telescopic equipment (i.e. pressure washers, brushes, roller, broom, etc.), but not physically;
2. Possibility of physical accessing to the element but only with heavy vehicles for the movement of people and only after removal of juxtapositions;
3. Possibility of physical accessing to the element but only with heavy vehicles for the movement of people;

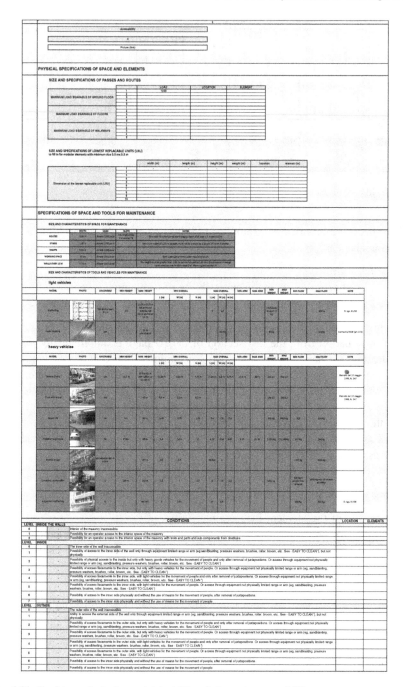

Fig. 4.11 Tool to evaluate the maintainability of elements: accessibility

4. Possibility of physical accessing to the element but only with light vehicles for the movement of people and only after removal of juxtapositions;
5. Possibility of physical accessing to the element but only with light vehicles for the movement of people;
6. Possibility of accessing to the building physically and without the use of means for the movement of people, after removal of juxtapositions;
7. Possibility of accessing to the building physically and without the use of means for the movement of people.

The external accessibility is repeated twice for the element external walls because in that case it is diversified between indoor and outdoor.

The final evaluation of accessibility is a value that ranges from 1 to 7 for the elements that are not needed for internal access and from 1 to 9 for element that has such a need.

Sheet h: final summary
The final page summarizes results of the evaluations given in the previous pages, with a synthetic value for each of the factors (Fig. 4.12).

4.4 A Set of Spreadsheets, to Evaluate the Uncertainty of Maintainability for All Technological Requirements of Each Given Typological Configuration of Technical Elements

A spreadsheet was prepared for all typological configurations of the technical elements in order to evaluate the degree of uncertainty over all their technological requirements (Fig. 4.13).

Each sheet is made up of a first page where all the technological requirement of the typological configuration of elements are listed (Fig. 4.14).

Then there are a number of other pages in each spreadsheet—one for each of the requirements listed in the first page—in which are reported all the interventions related to each of the specific technological requirements of the element (Fig. 4.15).

All interventions that have some influence on the technological requirements are mentioned and their statistical frequencies and specific requests for maintenance are

FACTORS OF MAINTAINABILITY	LEVEL
Accessibility	
Ergonomics	
Ease of assembly and disassembly	
Cleanability	
Substitutability	
Portability	
Visibility	

Fig. 4.12 Tool to evaluate the maintainability of elements: final summary

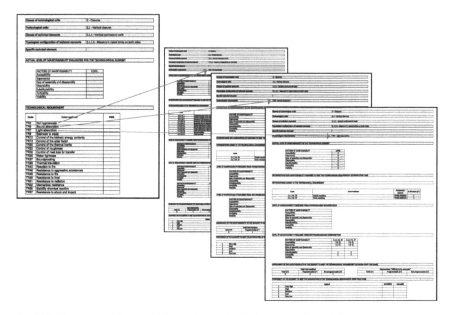

Fig. 4.13 Structure of the spreadsheet of a typological configuration of elements

reported. The frequencies came both from literature and from interviews with experts.

In these pages the level of maintainability that would be required for conducting each intervention is also reported, broken down for the 7 factors of maintainability (accessibility, ergonomics, ease of assembly and disassembly, cleanability, substitutability, portability, visibility). The level of maintainability estimated for each intervention was obtained using the *tool to evaluate the maintainability of elements*.

The set of spreadsheets presented above allows comparison of the level of maintainability required from each technological requirement of a typological configuration of technical elements with the actual level of maintainability of an element.

The corpus of tools and procedures introduced required a substantial amount of information to be created, therefore in order to realize them a number of databases and matrices of connections among various sources of information was created first (Fig. 4.16). All databases and matrices of connections were created, not to be fixed and untouchable, but as open tools implementable with feedback. Indeed, it is always possible to expand them by inserting or reviewing items.

Extracts from a few databases and matrices of correlation produced are reported below, as examples:

Database of interventions with frequencies
The database of interventions with frequencies (Fig. 4.17) was created both from interviews with experts and by a literature review (AA.VV. 1982, Albano 2005;

Classes of technological units	2 - Closures
Technological units	2.1 - Vertical closures
Classes of technical elements	2.1.1 - Vertical perimeteral walls
Typological configuration of technical elements	2.1.1.b - Masonry in naked bricks on both sides
Specific technical element	

ACTUAL LEVEL OF MAINTAINABILITY EVALUATED FOR THE TECHNOLOGICAL ELEMENT

FACTORS OF MAINTAINABILITY	LEVEL
Accessibility	
Ergonomics	
Ease of assembly and disassembly	
Cleanability	
Substitutability	
Portability	
Visibility	

TECHNOLOGICAL REQUIREMENT

Code	Requirements	Uncertinty
TR2	Not hygroscopic	
TR6	Sound absorption	
TR7	Light absorption	
TR9	Easiness to equip	
TR12	Control of the intrinsic energy contents	
TR13	Control of the solar factor	
TR20	Control of the thermal inertia	
TR23	Control of roughness	
TR28	Control of heat loss for transfer	
TR35	Water tightness	
TR37	Soundproofing	
TR39	Thermal insulation	
TR43	Reaction to fire	
TR46	Resistance to aggressive substances	
TR48	Resistance to fire	
TR49	Resistance to frost	
TR51	Resistance to radiation	
TR52	Mechanical resistance	
TR56	Stability chemical reactive	
TR57	Resistance to shock and impact	

Fig. 4.14 Sample of the first page of the spreadsheet

Bellini 1986; Croce 1994; Di Giulio 2003; Koenig et al. 1994; Lembo and Marino 2002; Mastrodicasa 1993; Nicolella 2003; Paganin 2003; Perret 1995).

Database of the maintainability required for interventions
The database of the level of maintainability required from all interventions (Fig. 4.18) was created by using the tool to evaluate the maintainability of elements. The use of the mentioned tool has allowed the translation of the conditions required for carrying out interventions into a number.

Matrix of connection between Typological configuration of technical elements/ Technological requirements and Interventions
The matrix of connection between each technological requirement (as listed in the Italian Standard UNI 8290-2:1983) of a typological configuration of technical

Classes of technological units	2 - Closures
Technological units	2.1 - Vertical closures
Classes of technical elements	2.1.1 - Vertical perimeteral walls
Typological configuration of technical elements	2.1.1.b - Masonry in naked bricks on both sides
Specific technical element	
Technological requirements	TR2 - Not hygroscopic

ACTUAL LEVEL OF MAINTAINABILITY OF THE TECHNOLOGICAL ELEMENT

FACTORS OF MAINTAINABILITY	LEVEL
Accessibility	0
Ergonomics	0
Ease of assembly and disassembly	0
Cleanability	0
Substitutability	0
Portability	0
Visibility	0

INTERVENTIONS AND MAINTAINABILITY REQUIRED TO KEEP THE TECHNOLOGICAL REQUIREMENT SATISFIED OVER TIME

INTERVENTIONS LINKED TO THE TECHNOLOGICAL REQUIREMENT

Code	Interventions		Frequency [years]	In 30 years [n°]
2.1.1 Int. 1	Visual inspection of the wall	indoor	1	30
2.1.1 Int. 2	Visual inspection of the wall	outdoor	1	30
2.1.1 Int. 9	Recovery of the damaged parts of the wall		WHEN NEEDED	/
2.1.1 Int. 16	Removing dirt and deposits on surface by chemical cleaning and washing	indoor	WHEN NEEDED	/
2.1.1 Int. 17	Removing dirt and deposits on surface by chemical cleaning and washing	outdoor	WHEN NEEDED	/
2.1.1 Int. 22	Partial replacement of elements	indoor	WHEN NEEDED	/
2.1.1 Int. 23	Partial replacement of elements	outdoor	WHEN NEEDED	/
2.1.1 Int. 24	Total replacement of elements	indoor	WHEN NEEDED	/
2.1.1 Int. 25	Total replacement of elements	outdoor	WHEN NEEDED	/
2.1.1 Int. 26	Partial remaking		WHEN NEEDED	/
2.1.1 Int. 27	Total remaking		WHEN NEEDED	/

LEVEL OF MAINTAINABILITY REQUIRED FROM PROGRAMABLE INTERVENTIONS

FACTORS OF MAINTAINABILITY	2.1.1 Int. 1	2.1.1 Int. 2
Accessibility		
Ergonomics		
Easy of assembly and disassembly		
Cleanability		
Substitutability		
Portability		
Visibility	4	4

LEVEL OF MAINTAINABILITY REQUIRED FROM NOT-PROGRAMMABLE INTERVENTIONS

FACTORS OF MAINTAINABILITY	2.1.1 Int. 6	2.1.1 Int. 16	2.1.1 Int. 17	2.1.1 Int. 23	2.1.1 Int. 24	2.1.1 Int. 25
Accessibility	4 (+ 2)	4 (+ 2)	4 (+ 2)	4 (+ 2)	4 (+ 2)	4 (+ 2)
Ergonomics	2 (+ 2)	2 (+ 2)	2 (+ 2)	3 (+ 2)	3 (+ 2)	3 (+ 2)
Easy of assembly and disassembly						
Cleanability	2			3		
Substitutability				2	2	2
Portability				2	2	2
Visibility						

FACTORS OF MAINTAINABILITY	2.1.1 Int. 26	2.1.1 Int. 27
Accessibility	4 (+ 2)	4 (+ 2)
Ergonomics	3 (+ 2)	3 (+ 2)
Easy of assembly and disassembly		
Cleanability		
Substitutability	2	2
Portability	2	2
Visibility		

ADEQUANCY OF THE MAINTAINABILITY OF THE ELEMENT TO KEEP THE TECHNOLOGICAL REQUIREMENT SATISFIED OVER TIME (AoM)

Total interventions			Interventions "difficult to be executed"		
Total [n°]	Programmable [n°]	Not-programmable [n°]	Total [n°]	Programmable [n°]	Not-programmable [n°]
11	2	9			

DEGREE OF UNCERTAINCY OVER THE SATISFACTION OF THE TECHNOLOGICAL REQUIREMENT OVER TIME

	Level	Distrib[%]	Cumu[%]
1	Very low		
2	Low		
3	Medium		
4	High		
5	Very high		

Fig. 4.15 Sample of a page of the spreadsheet relative to a specific technological requirement

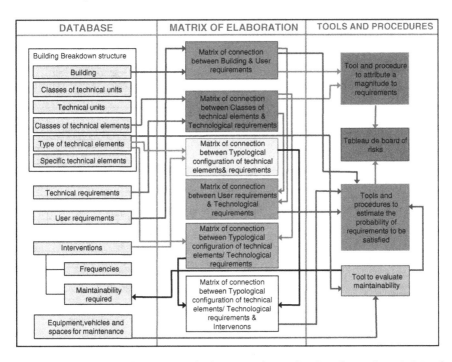

Fig. 4.16 Map of connections between databases, matrixes and tools and procedures (colour of arrows are agree to the box of destination)

CODE OF INTERVENTIONS	INTERVENTIONS	OPERATORS	FREQUENCY [years]	PROBABLE FREQUENCY [years]
2.1.1 Int. 1	Visual inspection of the wall I indoor	OPC	1	
2.1.1 Int. 2	Visual inspection of the wall I outdoor	OPC	1	
2.1.1 Int. 3	Inspection of the glass wall	OPC	1	
2.1.1 Int. 4	Check plates alignment	OPC	1	
2.1.1 Int. 5	Cleaning the coating surface of the wall I indoor	OPC	1	
2.1.1 Int. 6	Cleaning the coating surface of the wall I outdoor	OPC	WHEN NEEDED	5
2.1.1 Int. 7	Repainting I indoor	IMB	WHEN NEEDED	
2.1.1 Int. 8	Repainting I outdoor	IMB	2	
2.1.1 Int. 9	Recovery of the damaged parts of the wall	MRT	2	
2.1.1 Int. 10	Scraping and partial rebuilding of the plaster and / or paint I indoor	MRT	WHEN NEEDED	
2.1.1 Int. 11	Scraping and partial rebuilding of the plaster and / or paint I outdoor	MRT	WHEN NEEDED	20
2.1.1 Int. 12	Remaking of hooks	SRR	WHEN NEEDED	15
2.1.1 Int. 13	Tightening of the fixing elements for added coatings	SRR	WHEN NEEDED	10

Fig. 4.17 Extract from the database of interventions with frequencies

elements (Fig. 4.19) to the pertinent interventions was made by selecting from the list of all interventions of each typological configuration of technical elements, only those that play a role in maintaining each specific requirements. Indeed among a whole list of interventions that can be conducted on an element just a few are of some interest for specific technological requirements.

CLASSES OF TECHNICAL ELEMENTS	TYPE OF INTERVENTIONS	CODE OF INTERVENTIONS	INTERVENTIONS	FACTORS OF MAINTAINABILITY	LEVEL
		2.1.1 Int. 1	Visual inspection of the wall I indoor		
				Accessibility	
				Ergonomics	
				Easy of assembly and disassembly	
				Cleanability	
				Substitutability	
				Portability	
				Visibility	4
		2.1.1 Int. 2	Visual inspection of the wall I outdoor		
				Accessibility	
				Ergonomics	
				Easy of assembly and disassembly	
				Cleanability	
				Substitutability	
				Portability	
	Inspection			Visibility	4
		2.1.1 Int. 3	Inspection of the glass wall		
				Accessibility	
				Ergonomics	
				Easy of assembly and disassembly	
				Cleanability	
				Substitutability	
				Portability	
				Visibility	4
		2.1.1 Int. 4	Check plates alignment		
				Accessibility	4
				Ergonomics	
				Easy of assembly and disassembly	
				Cleanability	
External walls (vertical perimeteral walls)				Substitutability	
				Portability	
				Visibility	4

Fig. 4.18 Extract from the database of the maintainability required for interventions

Code	Typological configuration of technical elements	Code	Technological requirements	Code	Interventions
2.1.1.c	masonry in bricks with finishing in plaster on both sides				
		TR1	Reliability		
				2.1.1 Int. 1	Visual inspection of the wall I indoor
				2.1.1 Int. 2	Visual inspection of the wall I outdoor
				2.1.1 Int. 3	Inspection of the glass wall
				2.1.1 Int. 4	Check plates alignment
				2.1.1 Int. 14	Application of consolidation and protective treatments I indoor
				2.1.1 Int. 15	Application of consolidation and protective treatments I outdoor
				2.1.1 Int. 19	Removal of concrete and restoration of masonry subject to corrosion
				2.1.1 Int. 20	Restoring the missing parts
				2.1.1 Int. 21	Repair of concrete
				2.1.1 Int. 22	Partial replacement of elements I indoor
				2.1.1 Int. 23	Partial replacement of elements I outdoor
				2.1.1 Int. 26	Partial remaking
		TR2	Not hygroscopic		
				2.1.1 Int. 1	Visual inspection of the wall I indoor
				2.1.1 Int. 2	Visual inspection of the wall I outdoor
				2.1.1 Int. 3	Inspection of the glass wall
				2.1.1 Int. 4	Check plates alignment

Fig. 4.19 Extract from the matrix of connection between Typological configuration of technical elements/Technological requirements and Interventions

References

J.R. Albano, La maintenance des batiments, Paris, Le Moniteur, 2005. Trad. Ita. Talamo C. (ed.) *La manutenzione degli edifici.* (Sisemi editoriali, Napoli, 2005)

A. Bellini, *Tecniche della conservazione* (Franco Angeli, Milano, 1986)

G. Ciribini, Durabilità e problemi manutentivi nelle attività di recupero, in: Recuperare n.6, a.II, luglio-agosto (1983)

S. Croce, *Patologia edilizia; prevenzione e recupero*, in Manuale di progettazione edilizia, vol. 3 "Progetto tecnico e qualità", parte terza "La qualità nel tempo" (Hoepli, Milano, 1994)

S. Curcio, *Progettare la manutenzione*, in Manuale della progettazione edilizia, vol.3 (Hoepli, Milano, 1995)

S. Curcio, M. D'Alessandro (ed.), *Dalla manutenzione alla manutenibilità* (Angeli, Milano, 1994)

R. Di Giulio, *Manuale di manutenzione edilizia. Valutazione del degrado e programmazione della manutenzione* (Maggioli, Rimini, 2003)

International Code Council, *International Existing building Code of ICC*. (Country Club Hills, IL, 2003)

F. Kind Barkauskas, J. Brandt, S. Polònyi, B. Kauhsen, *Atlante del Cemento* (Utet Scienze Tecniche, Torino, 1998)

G.K. Koenig, B. Furiozzi, G. Fanelli, B. Bugatti, F. Brunetti, *Tecnologia delle costruzioni*, vol.III, (Le Monnier, Firenze, 1994)

F. Lembo, F.P.M. Marino, *Il comportmaneto nel tempo degli edifici* (EPC libri, Roma, 2002)

S. Mastrodicasa, *Dissesti statici delle strutture edilizie* (Editore Ulrico Hoepli, Milano, 1993)

C. Molinari, *La manutenzione programmata*, in Molinari C, Manuale di progettazione edilizia, vol. 3 «Progetto tecnico e qualità», parte terza «La qualità nel tempo» (Hoepli, Milano, 1994)

C. Molinari, *Il nuovo quadro di riferimento tecnico-normativo per la manutenzione*, in Curcio (ed) Manutenzione dei patrimoni immobiliari. Modelli, strumenti e servizi innovativi (Maggioli, Rimini, 1999)

C. Molinari, *Procedimenti e metodi per la manutenzione edilizia* (Esselibri editore, Napoli, 2002)

J. Natterer, T. Herzog, M. Volz, *Atlante del legno* (UTET Scienze tecniche, Torino, 1998)

M. Nicolella, *Programmazione degli interventi in edilizia. Guida al libretto di manutenzione del fabbricato* (UNI, Milano, 2003)

G. Paganin, *Danni e guasti dell'edificio. Professionisti tecnici e imprese* (Sistemi editoriali, Napoli, 2003)

J.Perret, *Guide de la maintenance des batiments*, Le Moniteur, Paris, 1995, trad.it.di C. Talamo, Guida alla manutenzione degli edifici, (Maggioli, Rimini, 2001)

C. Schittich, G. Staib, D. Balkow, W. Sobek, *Altlante del vetro* (UTET Scienze tecniche, Torino, 1998)

E. Schunck, T. Finke, R. Jenisch, H.J. Oster, *Atlante dei tetti* (UTET Scienze tecniche, Torino, 1998)

M. Zaffagnini, *Manuale di progettazione edilizia*, vol. 1 Tipologie e criteri di dimensionamento (Hoepli, Milano, 1994)

Standards and Laws

Ente Nazionale Italiano di Unificazione, 1981. UNI 8289:1981 Building. Functional requirements of final users

Ente Nazionale Italiano di Unificazione, 1981. UNI 8290-1 Edilizia residenziale. Sistema technologico. Classificazione e terminologia (Residential building. Building elements. Classification and terminology)

Ente Nazionale Italiano di Unificazione, 1983. UNI 8290-2:1983 Residential building. Building elements. Analysis of requirements

International Organization for Standardization, 1984. ISO 6241, Performance standards in building —Principles for their preparation and factors to be considered

International Organization for Standardization, 2009. ISO 31000:2009, Risk management— Principles and guidelines

Chapter 5
A Dashboard for Design Risk Management. Proposal for a Risk-Based Design Support and Lifelong Feedback Gathering System

Abstract The aim of this chapter is to propose a process to manage, at the design stage, the risks in buildings use and maintenance. The proposed process for risk management in architectural design consists of two steps: risk assessment and treatment and risk monitoring and reviewing. The first part uses the set of tools and methods introduced in Sect. 4.2 to attribute a degree of importance to an objective, during the brief phase, and a degree of uncertainty to them during the design stage. The combination of the two degrees defines the level of risk for all requirements that fill the dashboard of risks. Risks evaluated in this manner are managed according to the techniques for risk treatment from literature. The second part of the proposed process consists of a procedure for using feedback performance monitored during the phase of operation and management to review databases and evaluations. In this phase a method is presented for process monitoring that allows for constantly checking the actual performance of buildings and to compare them with those expected. Eventual mismatches between expected and actual performances are used in real-time updating of databases and evaluations.

Keywords Dashboard of risk · Degree of uncertainty · Degree of importance · Maintenability assessment · Feedback gathering · Databases implementation

5.1 Risk-Based Design Support and Lifelong Monitoring Tool: Framework

In this chapter a process is introduced to manage the risks over the objectives of buildings promoters and users with references to the phase of operation and management.

The risk management process aims, by the one side to support designers and promoters in evaluate design solutions with reference to the risks over objectives and, by the other, to help monitoring over the whole buildings life-time both performances and maintenance activities.

© The Author(s) 2015

C. Martani, *Risk Management in Architectural Design*,
PoliMI SpringerBriefs, DOI 10.1007/978-3-319-07449-8_5

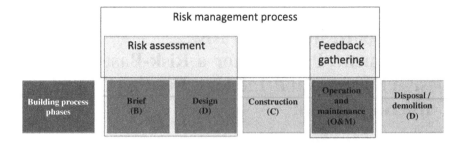

Fig. 5.1 Phases of the process of risk management over long-term objectives

To reach this purpose the process is divided in two phases: risk assessment and treatment, to evaluate project with reference to risks, and risk monitoring and reviewing to gather feedback information from the phase of operation and maintenance (Fig. 5.1).

5.1.1 Risk Assessment and Treatment

Risk assessment aims to produce a risk-based design support, useful both for designers to realize project more oriented to promoters' and users' needs, and for promoters to evaluate design solutions for their likelihood to meet declared objectives.

In line with these aims the proposed risk assessment introduces a dashboard of risks in 3 steps, each of which with specific actions and goals (see Fig. 5.2). The 3 steps are:

- Risk identification. At this step a level of importance is given to all requirements that represent promoters and users goals in order to point those objective that are "hazards" (possible sources of undesired events).
- Risk analysis. At this step a degree of uncertainty is given to all requirements through the definition of both, the need for maintenance of the requirements, and the maintainability of elements they rely on.
- Risk evaluation and treatment. At this step risk are estimated and represented into a dashboard of risks to be managed.

5.1.2 Risk Reviewing Through Feedback Information

The action of monitoring—which is made of feedbacks gathering in the operation and management phase—describes the actual condition of buildings in use, while the action of reviewing consists in all change in databases and in previsions, following up with what feedbacks suggest.

The present work proposes the use of feedbacks to correct prevision on risks and databases and to improve the effectiveness of future designs (Fig. 5.3).

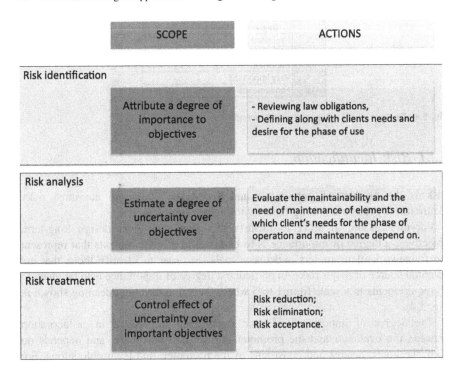

Fig. 5.2 Scope and actions of the steps of risk assessment on design proposal

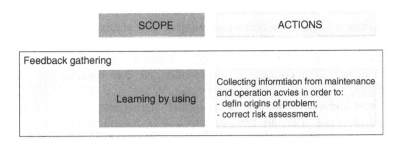

Fig. 5.3 Scope and actions of risk reviewing through feedback information from the use and maintenance

5.2 Risk Assessment: A Risk-Based Design Support

In the present work the phase of risk assessment aims to evaluate design proposal in order to support designers and promoters to identify the major risks over a framework of given long-term goals.

The risk assessment process proposed is organized along three steps: risk identification, risk analysis and risk evaluation and treatment.

Degree of importance	
1	Irrelevant
2	Of little importance
3	Important
4	Very important
5	Fundamental

Fig. 5.4 Degrees of importance of requirements

5.2.1 Risk Identification

Risk is the product of importance and uncertainty, therefore the most risky requirements are settle among those with highest importance.

For this reason, in order to identify the main risks over design long-term objectives, a degree of importance is to be given to all requirements that represent the framework of promoters and users goals, in order to identify those that are potential source of major risks. In particular a degree of importance is to be given to all requirements in a scale from 1 to 5 where the value have the meaning shown in Fig. 5.4.

The degree of importance of each requirement is defined in collaboration between the evaluator and the promoter during the brief phase, and depends on both: the goals, desires and constrains of the promoter, and legal obligations. For instance a requirement that is neither mandatory by low, nor particularly relevant for the promoter will be given a degree of importance equal to 1 (irrelevant). Instead a requirement that is either mandatory by low, or very relevant for the promoter will be given a degree of importance equal to 5 (fundamental).

The output of risk identification consists of two matrixes: one for the user requirements, in which a degree of importance is assigned to all requirements of each class of technical elements (Fig. 5.5), and one for the user requirements, in which a degree of importance is attributed to all user requirements as well (Fig. 5.6).

The most important requirements are named hazards, meaning with that, possible sources of risk. They are those on which to assess and treat the risk.

5.2.2 Risk Analysis

Risk analysis is to be conducted either at the design stage, or after design. The aim of risk analysis is to evaluate the degree of uncertainty around requirements. Indeed, in order to assess the level of risk over the hazards previously identified it is necessary to estimate their uncertainty first. The procedure for estimating the uncertainty around the most important requirements is distinguished in two types: a procedure for technological requirements and one for user requirements.

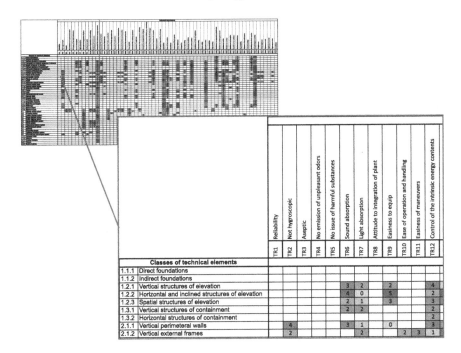

Fig. 5.5 Matrix of the degree of importance of technical requirements

Classes of technical elements	TR1 Reliability	TR2 Not hygroscopic	TR3 Aseptic	TR4 No emission of unpleasant odors	TR5 No issue of harmful substances	TR6 Sound absorption	TR7 Light absorption	TR8 Attitude to integration of plant	TR9 Easiness to equip	TR10 Ease of operation and handling	TR11 Easiness of maneuvers	TR12 Control of the intrinsic energy contents
1.1.1 Direct foundations												
1.1.2 Indirect foundations												
1.2.1 Vertical structures of elevation							3	2	2			4
1.2.2 Horizontal and inclined structures of elevation							4	0	5			2
1.2.3 Spatial structures of elevation							2	1	3			3
1.3.1 Vertical structures of containment							2	2				2
1.3.2 Horizontal structures of containment												2
2.1.1 Vertical perimeteral walls	4						3	1	0			3
2.1.2 Vertical external frames	2							2		2	3	1

Fig. 5.6 Matrix of the degree of importance of user requirements

Building and building areas	UR1.1 Morphological efficiency in relation to actions (both static and dynamic)	UR2.1 Evacuation in case of emergency	UR2.2 Control of risk of fire	UR3.1 Safety against pollutants	UR4.1 Protection against falling	UR5.1 Security against intrusion	UR6.1 Control of internal air temperature	UR6.2 Summer comfort	UR6.3 Control of orientation	UR7.1 Control of noise	UR8.1 Control of natural lighting	UR9.1 Control of ventilation	UR9.2 Control of odors	UR10.1 Resistance to external agents	UR10.2 Resistance to the stresses of exercise	UR11.1 Accessibility to people and things	UR11.2 Accessibility for people with disabilities	UR12.1 Furnishability	UR12.2 Practicability	UR13.1 Adaptability over time (possibility of merging, splitting, spaces)	UR14.1 Functional integration	UR14.2 Dimensional integration	UR15.1 Control of pollution	UR16.1 Energy efficiency and use of natural energy
The whole building	5	5	5	4	4	3	4	4	4	4	4	3	3	2	3	5	4	2	3	3	2	1	2	2
Area 1	5	5	5	3	5	4	2	2	4	5	4	3	3	2	5	4	2	3	1	1	1	2	2	
Area 2	5	5	5	5	3	4	4	4	4	3	3	3	3	2	4	5	4	2	3	5	3	1	2	2

5.2.3 Procedure for Estimate the Uncertainty Around Technical Requirements

The first operation to be conducted for estimate the uncertainty around technical requirements is to define all specific technical elements within a given design project that can be tracked down to the classes of technical elements on which risks were identified (Fig. 5.7).

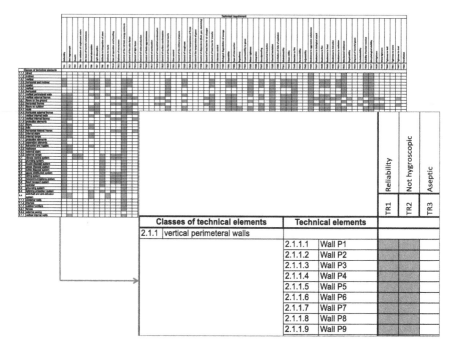

Fig. 5.7 Map of connections between technological requirements and specific technical elements based on which the uncertainty can be estimated

The satisfaction over time of the technological requirements of a technical element largely depends on predisposition of the element to keep its performances over time. Within the building process the performances required to satisfy over time the objectives in the operation and management phase are very often dependent to the maintenance activities, that are the more likely to be carried out, the more the conditions needed for their conductions are guaranteed.

Therefor the proposed procedure aims to verify if the conditions needed to deliver the interventions required to keep requirements satisfied over time are given. The highest is the number of interventions that cannot be carried out properly, the highest is the uncertainty around the satisfaction over time of the specific requirement.

To estimate the uncertainty around technical requirements the procedures proposed 3 steps:

1. evaluation of the maintainability of the element;
2. comparison between maintainability require from technological requirements and the actual maintainability of elements;
3. estimation of the uncertainty around the ability of an element to keep the satisfaction of its technological requirements over time.

5.2.4 Evaluation of the Maintainability of the Element

Using the tool to evaluate the maintainability, introduced in Chap. 4, the level of maintainability of an element can be assessed with reference to 7 factors: accessibility, ergonomics, easy of assembly and disassembly, cleanability, substitutability, portability and visibility.

To this purpose it is necessary first to collect all design documents of the building to evaluate. The evaluation begins from the elements that have one or more technical requirements with a degree of importance of 5 (i.e. in Fig. 5.8 an example is reported in which fundamental importance was assigned to the requirement of Not hygroscopic of vertical perimeter walls).

To this purpose the tool to evaluate maintainability of the pertinent element is chosen and a level is given to all 7 factors of maintainability based on the features of the element and of spaces around the it.

In Fig. 5.9 an example of fulfilment of two factors of maintainability is reported, where to the factor of accessibility is given a level of 4, which means that there is the possibility of physical access to the element but only with light vehicles for the movement of people and only after removal of juxtapositions. While to the factor of ergonomics a level of 5 is given, "which means that space next to the element can accommodate a person and enable him to operate with bulky equipment (vacuum cleaners, pressure washers, sanders, welders, etc.) and to elements of manoeuvre and components to be replaced", and that it is possible to inspect the interior of the element.

At the end of the operation a final summary of the evaluation derived from the tool in which a level is given to all factors (Fig. 5.10).

Classes of technical elements		Reliability	Not hygroscopic	Aseptic	No emission of unpleasant odors	No emission of harmful substances	Sound absorption
		TR1	TR2	TR3	TR4	TR5	TR6
2.1.1	Vertical perimeteral walls		5				3
2.1.2	Vertical external frames						3

Fig. 5.8 Example of technological requirements with high importance

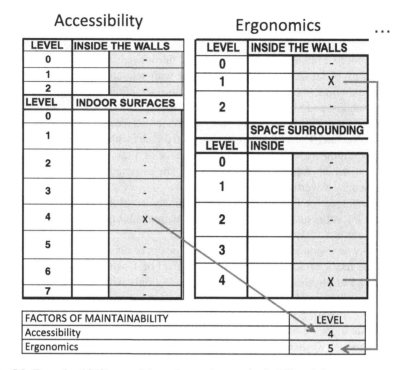

Fig. 5.9 Example of fulfilment of the tool to evaluate maintainability of elements

FACTORS OF MAINTAINABILITY	LEVEL
Accessibility	4
Ergonomics	5
Ease of assembly and disassembly	3
Cleanability	3
Substitutability	1
Portability	2
Visibility	4

Fig. 5.10 Final summary of the maintainability evaluation for an element

5.2.5 Comparison Between Maintainability Require from Technological Requirements and the Actual Maintainability of Elements

The final summary of the tool to evaluate the maintainability serves as an input information to fulfil the sheet for the estimation of the uncertainty over all technological requirements of each element.

The sheets prepared for the requirements of each typological configuration of technical elements reports the maintainability required from all interventions necessary to maintain the element with references to the requirements. Indeed, not all requirements of a given element are influenced from the same number or type of interventions. For instance, among all interventions of maintenance that can be conducted on masonry in naked bricks, those that are of some relevance for the requirement of sounds insulation are almost zero, while those pertinent to the requirement of not hygroscopic are up to 11. In this sense the requirement of not hygroscopic is more maintenance dependent than that of sounds insulation.

Therefore in order to estimate the uncertainty over all technological requirements of an element the output of the evaluation of maintainability of the element is reported into the pertinent sheet and is compared with the maintainability required from all requirements (Fig. 5.11).

5.2.6 Estimation of the Uncertainty Around the Ability of an Element to Keep the Satisfaction of Its Technological Requirements Over Time

In this way the level of adequacy of maintenance (AoM) is estimated (Fig. 5.12) as the number of interventions, both programmable and not programmable, that are "difficult to perform" or, in other words, the maintainability of which is inadequate with reference to the maintainability requested.

Once the adequacy of maintainability of each given requirement is defined and, along with that, its propensity to be satisfied over time is also forecasted, the degree of uncertainty of the requirement can be estimated. Indeed, a degree of uncertainty can be given to each technological requirement based on the number of interventions "difficult to perform" (the maintainability of which is inadequate), to be conducted in a slot time.

The number of interventions to be conducted in a slot of time is computed as the number of programmable interventions, times the number of time they are expected to be conducted within a period. For instance an intervention with annual frequency is considered to be conducted 30 times in a time horizon of 30 years.

Five possible levels of uncertainty of requirements have been defined: very low, low, medium, high and very high. The level is given on the bases of the subdivision shown in Fig. 5.13.

In the example reported, for instance, the adequacy of maintainability of the technological requirement of *Not hygroscopic* is *very low* (1), as there aren't programmable interventions "difficult to perform".

The degree of uncertainty found in this way is a single, most likely, value that indicates the propensity of a technological requirement to be satisfied over time. But as the degree of confidence on predictions can't be as strong as a deterministic value express, in order to carry out future statistical computation the level of uncertainty

Classes of technological units	2 - Closures

Technological units	2.1 - Vertical closures

Classes of technical elements	2.1.1 - Vertical perimeteral walls

Typological configuration of technical elements	2.1.1.b - Masonry in naked bricks on both sides

Specific a technical element	EW_02

Technological requirements	TR2 - Not hygroscopic

ACTUAL LEVEL OF MAINTAINABILITY OF THE TECHNOLOGICAL ELEMENT

FACTORS OF MAINTAINABILITY	LEVEL
Accessibility	4
Ergonomics	5
Easy of assembly and disassembly	3
Cleanability	3
Substitutability	1
Portability	2
Visibility	4

INTERVENTIONS AND MAINTAINABILITY REQUIRED TO KEEP THE TECHNOLOGICAL REQUIREMENT SATISFIED OVER TIME

INTERVENTIONS LINKED TO THE TECHNOLOGICAL REQUIREMENT

Code	Interventions	Frequency [years]	In 30 years [n°]
2.1.1Int. 1	Visual inspection of the wall I indoor	1	30
2.1.1Int. 2	Visual inspection of the wall I outdoor	1	30
2.1.1Int. 9	Recovery of the damaged partsof the wall	WHEN NEEDED	/
2.1.1Int. 16	Removing dirt and deposits on surfaceby chemical cleaning and washing I indoor	WHEN NEEDED	/
2.1.1Int. 17	Removing dirt and deposits on surface by chemical cleaning and washing I outdoor	WHEN NEEDED	/
2.1.1Int. 22	Partial replacement of elements I indoor	WHEN NEEDED	/
2.1.1Int. 23	Partial replacement of elements I outdoor	WHEN NEEDED	/
2.1.1Int. 24	Total replacement of elements I indoor	WHEN NEEDED	/
2.1.1Int. 25	Total replacement of elements I outdoor	WHEN NEEDED	/
2.1.1Int. 26	Partial remaking	WHEN NEEDED	/
2.1.1Int. 27	Total remaking	WHEN NEEDED	/

LEVEL OF MAINTAINABILITY REQUIRED FROM PROGRAMMABLE INTERVENTIONS

FACTORS OF MAINTAINABILITY	2.1.1Int. 1	2.1.1Int. 2
Accessibility		
Ergonomics		
Easy of assembly and disassembly		
Cleanability		
Substitutability		
Portability		
Visibility	4	4

LEVEL OF MAINTAINABILITY REQUIRED FROM NOT-PROGRAMMABLE INTERVENTIONS

FACTORS OF MAINTAINABILITY	2.1.1Int. 9	2.1.1Int. 16	2.1.1 Int. 17	2.1.1Int. 22	2.1.1 Int. 23	2.1.1 Int. 24	2.1.1 Int. 25
Accessibility	4 (+2)	4 (+2)	4 (+2)	4 (+2)	4 (+2)	4 (+2)	4 (+2)
Ergonomics	2 (+2)	2 (+2)	2 (+2)	2 (+2)	3 (+2)	3 (+2)	3 (+2)
Easy of assembly and disassembly							
Cleanability	2		3				
Substitutability				2	3	2	3
Portability				2	2	2	2
Visibility							

FACTORS OF MAINTAINABILITY	2.1.1Int. 26	2.1.1 Int. 27
Accessibility	4 (+2)	4 (+2)
Ergonomics	3 (+2)	3 (+2)
Easy of assembly and disassembly		
Cleanability		
Substitutability	2	2
Portability	2	2
Visibility		

Fig. 5.11 Section concerning the level of maintainability required from interventions, within the sheet of the technological requirement of not hygroscopic of a specific brick wall

Interventions"difficult to be executed"		
Total [n°]	Programmable [n°]	Not-programmable [n°]
6	0	6

Fig. 5.12 Section concerning adequacy of maintainability, within the sheet of the technological requirement of not hygroscopic of a specific brick wall

	Degree of uncertainty	Number of programmable interventions "difficult to be executed"
1	Very low	0
2	Low	0<X<7
3	Medium	7<X<15
4	High	15<X<31
5	Very high	31+

Fig. 5.13 Range to attribute the degrees of uncertainty

over technological requirements is better represented (more informative) as a probability distribution, rather than a single value.

For this reason a set of rules have also been settle to define a probability distribution around the most likely level of uncertainty (that have as mean the deterministic value identified as above), on the basis of the number of the not-programmable interventions that are "difficult to perform".

The adequacy of maintainability of not programmable interventions hasn't been considered in establishing the mean value of the degree of uncertainty because those are interventions that may never be conducted on the service life of an element. But they are not irrelevant, because even if not programmable are still possible and therefor the many of them are "difficult to perform", the less the requirements that rely on them are likely to be guarantee.

In the example reported, for instance, the probability distribution around the degrees of uncertainty for the technological requirement "Not hygroscopic" is distributed along values 1 (65 %) and 2 (35 %), as the most likely level is the optimal one but there are 3 not programmable interventions "difficult to perform" (Fig. 5.14).

Once the sheet of an element is completed, a degree of uncertainty and a probability distribution around it is given for all relative requirements.

The matrix of uncertainty over technological requirements can then be fulfilled with the degrees of uncertainty derived from all sheets of the element interested (Fig. 5.15).

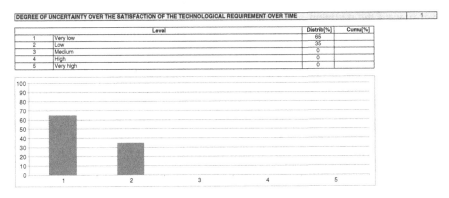

Fig. 5.14 Section concerning the evaluation of the degree of uncertainty of the technological requirement "not hygroscopic" of a specific brick wall, and its distribution

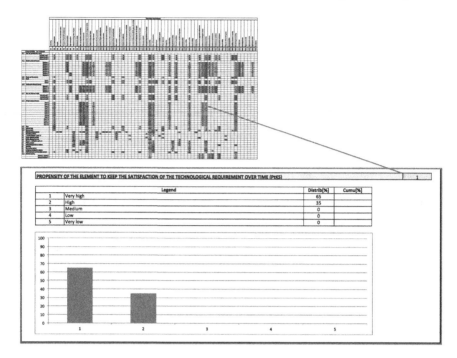

Fig. 5.15 Representation of the connection between all cells of the matrix of uncertainty of technological requirements and the relative sheet

5.2.7 Procedure for Estimate the Uncertainty Around User Requirements

User requirements are requirements that depend on the performances of a number of technical elements, therefore the uncertainty around those requirements is strongly influenced by the propensity of all elements involved to guarantee the satisfaction over time of the pertinent requirements.

To estimate the uncertainty of user requirements over time a procedure is presented in 3 steps:

1. Identification of all elements, and pertinent technological requirements, involved with the user requirements and evaluation of their uncertainty;
2. Definition of conditions to run the Monte Carlo simulations;
3. Estimation of the level of uncertainty around the user requirement through simulations.

5.2.8   Identification of All Elements, and Relative Technological Requirements, Involved with a User Requirement

The first step to estimate the uncertainty around user requirements is to identify elements, and pertinent technological requirements, involved with a user requirement is to select the technological requirements of interest for the user requirement on which the risk analysis is being conducting by using the map of connections between user requirements and technological requirements. Then select all classes of technical elements that are associated to the technological requirement.

And, finally, select all specific elements of the building under evaluation that according to the previous filter are related to user requirement. In this way a map of all technological requirements of elements involved with a user requirement would be created (Fig. 5.16).

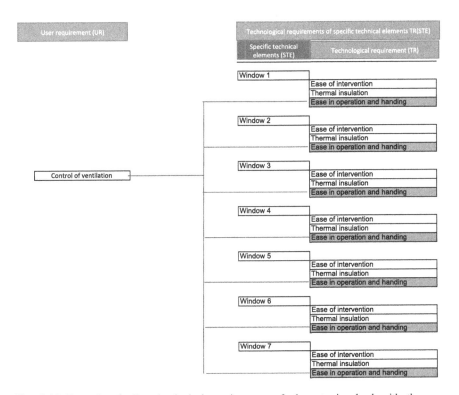

Fig. 5.16 Example of all technological requirements of elements involved with the user requirement of "control of ventilation" in a given building

5.2.9 Definition of Conditions to Run the Monte Carlo Simulation

All technological requirements of elements involved with the user requirement to be evaluated have already been evaluated (see point 1) therefore all of them are associated to sheets where their uncertainty are reported, both as a single value, and as a probability distribution (Fig. 5.17).

With these information the Monte Carlo simulation can be run. The first step to set the Monte Carlo simulations is to prepare a table with all technological requirements of elements involved and, for all of them, the cumulative probabilities of their level of uncertainty (Fig. 5.18) have to be reported.

Then a simulation can be settled. To this purpose for all technological requirements that stand in the table a random number has to be chosen in a range from 1 to 100[1] and a level of uncertainty has to be associated to the random number extracted in accordance to the probability distribution of the pertinent requirement.

Indeed, looking at the example of Fig. 5.19 it can be noticed that if the degree uncertainty of a technological requirement has the following probability distribution

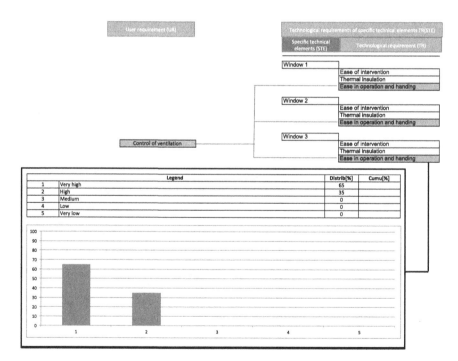

Fig. 5.17 Example of all technological requirements of elements involved with the user requirement of control

---

[1]   With the formula "=RAND()*100" Microsoft Excel does that automatically.

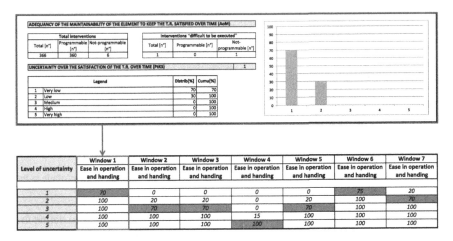

Fig. 5.18 All element involved in the simulations, with pertinent technological requirement and the cumulative probability of uncertainty

along 5 values: level 1 = 70 %; level 2 = 30 %; levels 3, 4 and 5 = 0 %, then the probability that a number from 1 to 70 has to be extracted in a random choice of 100 numbers is the same probability that the level 1 has to be the more appropriate degree of uncertainty. The same way, the probability that a number from 71 to 100 have to be extracted in a random choice from 100 numbers is the same probability that the level 2 has to be the more appropriate degree of uncertainty. For this reason the level of uncertainty selected according to random samples can be trust to represent a what is expected to happen in reality.

By repeating the same operation for all technological requirements of the elements involved with the user requirement a full row is prepared for the simulation of the overall expected uncertainty.

Fig. 5.19 Sample of attribution of a level of uncertainty to ta random number, in accordance to a probability distribution

	B	C
8		Window 1
9	Level of uncertainty	Ease in operation and handing
10		
11	1	70
12	2	100
13	3	100
14	4	100
15	5	100
16		
17	Random number	54	← = RAND()*100
18	Level of uncertainty	1

= IF(C17> 70;"2";IF(C17< 70;"1";IF(C17= 70;"1")))

```
=IF(R18>1;"5";IF(Q18>1;"4";IF(P18>1;"3";IF(O18>1;"2";IF(N18>1;"1")))))
```

Level of uncertainty	Window 1 Ease in operation and handing	Window 2 Ease in operation and handing	Window 3 Ease in operation and handing	Window 4 Ease in operation and handing	Window 5 Ease in operation and handing	Window 6 Ease in operation and handing	Window 7 Ease in operation and handing
1	70	0	0	0	0	75	20
2	100	20	20	0	20	100	70
3	100	70	70	0	70	100	100
4	100	100	100	15	100	100	100
5	100	100	100	100	100	100	100
Random number	39	64	39	32	56	48	12
Level of uncertainty	1	3	3	5	3	1	1	3

Fig. 5.20 Example of conditions settlement to determine the level of uncertainty to attribute to the whole user requirement

Finally the last aspect to be settled in order to run the simulations is to define the conditions that determine the level of uncertainty to attribute to the whole user requirement depending on the configuration of all levels of uncertainty involved. In the example of Fig. 5.20 a solution was proposed in which: the couple of elements with the highest degree of uncertainty define the degree of uncertainty of the whole system.

This is just one possible solution, among many, that seems reasonable for the specificity of the project. But conditions that settle the overall uncertainty of a user requirement have to be defined project by project, along with clients.

5.2.10 Estimation of the Level of Uncertainty Around the User Requirement Through Simulations

Once all technical elements, with pertinent technological requirements involved with the user requirement, are identified and all conditions are settled then an appropriate number of simulations (that range from 1000 to 5000) can be run and results collected.

The distribution of results, in percentage, indicates the level of uncertainty of the whole user requirement.

Finally the degree of uncertainty obtained this way goes into the matrix of uncertainty of user requirements (Fig. 5.21), along with all degrees of uncertainty of the other user requirements.

5.2.11 Risk Evaluations and Treatment

Risk evaluations is the last step of the risk assessment process that aims to produce a dashboard of risks, where all requirements, both technological and user requirements, are associated to a level of risk.

Building and building areas	UR1.1	UR2.1	UR2.2	UR3.1	UR4.1	UR5.1	UR6.1	UR6.2	UR6.3	UR7.1	UR8.1	UR9.1	UR9.2	UR10.1	UR10.2	UR11.1	UR11.2	UR12.1	UR12.2	UR13.1	UR14.1	UR14.2	UR15.1	UR16.1
The whole building	3	2	3	2	2	3	2	3	2	3	2	4	2	3	3	2	2	3	3	3	1	2	4	4
Area 1	4	2	3	1	2	4	2	2	1	5	1	5	1	4	2	1	1	2	5	1	1	1	4	4
Area 2	3	2	2	3	3	3	3	4	4	2	2	3	3	3	5	4	2	4	1	5	2	2	5	5

User requirements column legend (left to right): Morphological efficiency in relation to actions (both static and dynamic); Evacuation in case of emergency; Control of risk of fire; Safety against pollutants; Protection against falling; Security against intrusion; Control of internal air temperature; Summer comfort; Control of orientation; Control of noise; Control of natural lighting; Control of ventilation; Control of odors; Resistance to external agents; Resistance to the stresses of exercise; Accessibility to people and things; Accessibility for people with disabilities; Furnishability; Practicability; Adaptability over time (possibility of merging, splitting, spaces); Functional integration; Dimensional integration; Control of pollution; Energy efficiency and use of natural energy

Level of uncertainty	Window 1 Ease in operation and handing	Window 2 Ease in operation and handing	Window 3 Ease in operation and handing	Window 4 Ease in operation and handing	Window 5 Ease in operation and handing	Window 6 Ease in operation and handing	Window 7 Ease in operation and handing	OUTPUT OF SIMULATION [n]	OUTPUT OF SIMULATION [%]
								1000	
1	70	0	0	0	0	75	20	18	2
2	100	20	20	0	20	100	70	198	20
3	100	70	70	0	70	100	100	501	50
4	100	100	100	15	100	100	100	283	28
5	100	100	100	100	100	100	100	0	0

Fig. 5.21 Representation of the connection between all cells of the matrix of uncertainty of user requirements and the relative sheet of simulation (see Sect. 5.2)

To this purpose, according with the definition of risk given by the International Standard ISO31000, the risk of all requirements is computed (see Fig. 5.22) by multiplying the degree of importance (that range from 1 to 5), and the degree of uncertainty (that range from 1 to 5 as well).

Fig. 5.22 Computation of the risk level of the dashboard

Importance

Classes of technical elements		Reliability (TR1)	Not hygroscopic (TR2)
2.1.1	Vertical perimeteral walls		5

X

Uncertainty

Classes of technical elements			
2.1.1	Vertical perimeteral walls		
		External wall 1	2
		External wall 2	4
		External wall 3	1

=

Risk

Classes of technical elements			
2.1.1	Vertical perimeteral walls		
		External wall 1	10
		External wall 2	20
		External wall 3	5

The risk obtained in this way (in a range from 0 to 25) is reported into the two matrixes of the dashboard of risks: the dashboard of technical requirements (Fig. 5.23) and the dashboard of user requirements (Fig. 5.24).

For each requirement within the dashboard the pertinent sheet can be consulted in order to track the level of risk down to its origins and to treat the risk in accordance to the strategies of risk treatment presented into the International Standard ISO31010.

5.3 Risk Reviewing Through Feedback Information: Proposal of a Method to Gather Real Time Information for Correcting Previsions and Improving Future Design

Feedbacks from use are the information gathered in the operation and management phase that describe the actual condition of buildings in use. Reviewing consists in all change in database and in previsions, following up with feedback suggestions.

Feedbacks information from the phase of operation and management are precious for the risk management process because can help to both: don't repeat the same mistakes on future project designs and correct problems on operating buildings, if appropriate.

Within the present work feedbacks are the actual data recorded from activities of conduction. The figures involved in feedback gathering activities are operators of the companies of cleaning and maintenance. Indeed within the proposed framework operators have to report into a database all interventions delivered and for each of them they have to specify: data, duration, costs and tools and accessory used, other than eventual difficulty in conducting activities.

Feedbacks from operation and management lead to the section of reviewing and updating of both: databases and estimations of the degree of uncertainty over requirements.

With reference to timing, data are gathered every time operators conduct an intervention. While review and updates of database and estimations are conducted periodically, on the base of promoter' and users' needs. Databases that can be reviewed following up with feedbacks are three:

- database of level of maintainability;
- database of interventions (Fig. 5.25);
- database of frequency of interventions.

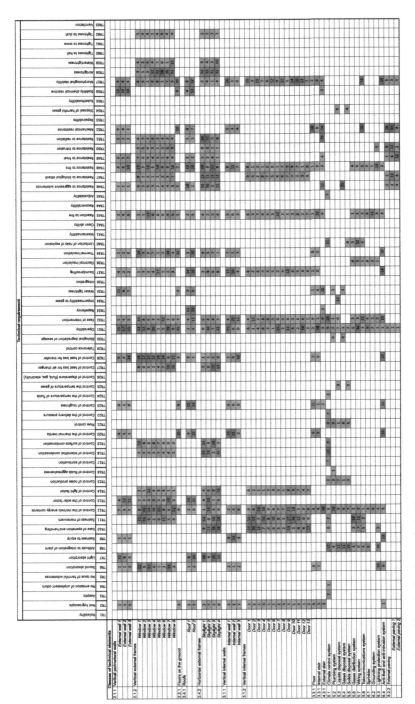

Fig. 5.23 Dashboard of technological risks

Building and building areas																							
	UR1.1	UR2.1	UR2.2	UR3.1	UR4.1	UR5.1	UR6.1	UR6.2	UR6.3	UR7.1	UR8.1	UR9.1	UR9.2	UR10.1	UR10.2	UR11.1	UR11.2	UR12.1	UR12.2	UR13.1	UR14.1	UR14.2	UR15.1	UR16.1
The whole building		10	14	8	10	14	7	9	10	14	7	12	6	7	10	11	7	6	9	9	2	2	9	9
Area 1	10	10	15	3	10	16	4	4	4	25	6	14	4	8	4	5	4	4	15	1	1	1	8	8
Area 2	10	11	12	15	8	12	13	15	6	6	10	8	6	10	16	10	9	4	25	5	2	10	10	

Fig. 5.24 Dashboard of the risk of user requirements

CODE OF INTERVENTIONS	INTERVENTIONS	OPERATOR	FREQUENCY
3.2.1 Int. 1	Inspection and verification of the conditions I floor	OPC	0.5
3.2.1 Int. 2	Inspection and verification of the conditions I celing	OPC	3
3.2.1 Int. 3	Analysis of the degradation of the wooden elements I celing	TSP	7
3.2.1 Int. 4	Cleaning of the floor	OPC	0.1
3.2.1 Int. 5	Deep washing of the floor	OPC	0.3
3.2.1 Int. 6	Washing the floor with tools or machines (i.e. Rotowash)	OPC	0.3
3.2.1 Int. 7	Remove crusts through mechanical abrasion	OPC	WHEN NEEDED
3.2.1 Int. 8	Remaking of the painting of the celing	IMB	7

CODE OF INTERVENTIONS	INTERVENTIONS	OPERATOR	FREQUENCY
3.2.1 Int. 1	Inspection and verification of the conditions I floor	OPC	0.5
3.2.1 Int. 2	Inspection and verification of the conditions I celing	OPC	3
3.2.1 Int. 3	Analysis of the degradation of the wooden elements I celing	TSP	7
3.2.1 Int. 4	Cleaning of the floor	OPC	0.1
3.2.1 Int. 33	Chemical cleaning of floor	OPC	1
3.2.1 Int. 5	Deep washing of the floor	OPC	0.3
3.2.1 Int. 6	Washing the floor with tools or machines (i.e. Rotowash)	OPC	0.3
3.2.1 Int. 7	Remove crusts through mechanical abrasion	OPC	WHEN NEEDED
3.2.1 Int. 8	Remaking of the painting of the celing	IMB	7

Fig. 5.25 Review of the number and type of interventions for element, based on feedbacks

Moreover on the bases of the review on database also estimations for risk assessment need to be updated. Indeed estimations of the degree of uncertainty are based on number and frequencies of interventions, other than on the level of maintainability required. Therefore when premises change consistently, also the estimations of uncertainty have to be updated.

Part III
Applications: Tests on Case Studies

Chapter 6
Application of the Dashboard for Risk Management: The Case Study of Two Buildings of Worship

Abstract The aim of this chapter is to test on real world case studies the set of tools and methods to create a dashboard that has been proposed in the previous chapters. Tests were run on two buildings of worship. To this purpose a degree of importance was assigned to a set of requirements, both user and technological, that represent the needs of the Italian Council of Bishops (CEI) with reference to the phase of use of churches. Then a degree of importance was estimated for all requirements by using spreadsheets and Monte Carlo simulations. Results show that the two churches generally have a low level of risk related to the use and maintenance, but that some particular elements, such as windows and roofs, can bring serious problems over time, with reference to the requirements of "control of roughness" of "control of ventilations". Main results are presented and commented upon.

Keywords Applications · Case studies · Building commissioning · Buildings of worship · Churches Design · Churches Maintenances · Database Implementation

6.1 The Objectives of the Italian Council of Bishops for the Buildings of Worship

Within the context of research collaboration between the ABC department of the Politecnico di Milano and the Italian Council of Bishops (CEI), an opportunity arose to test the risk management process for buildings long-term objectives on real word case studies.

Indeed two buildings of worship were used as cases study for testing the risk management process. As introduced in Chaps. 4 and 5, risk management process is organized along two phases: the phase of risk assessment, that aims to identify, analyse and treat risks right at the design stage, and the phase of feedback gathering that is finalized to learn from experiences reached in use and maintenance.

Within this framework, the first step of the process is the risk assessment. Therefore, according to the proposed procedure, the first operation to be done is the

© The Author(s) 2015
C. Martani, *Risk Management in Architectural Design*,
PoliMI SpringerBriefs, DOI 10.1007/978-3-319-07449-8_6

definition of objectives and the attribution of a level of importance to all requirements that represent these objectives.

With reference to buildings of worship the definition of objectives were conducted considering CEI as the promoter of the constructions of new churches.

CEI is neither the final user, nor the owner of the largest part of the buildings of worship in Italy. Indeed most of the churches are promoted, constructed, owned and managed by the communities and the responsibility for these activities is with the parishes. The final owners of churches are the territorial communities where churches are built and used. Therefore also the objectives for the phase of operation and maintenance of churches have to be defined by the communities and by the responsible parishes.

Nevertheless the indications of the CEI for the construction and the restoration of buildings of worship are important for parishes in order to define needs in the phase of commissioning the construction of new churches. To some extent it can be considered that indications from CEI serve as a platform on which each community built the frame of its own specific needs. Therefor it is reasonable to take the set of goals and needs expressed by CEI as the broad and general set of objectives for all churches.

CEI's office for building of worship[1] has produced a series of documents that serve as an indication for the construction and restoration of buildings of worship. These documents are available at the website of the office and can be accessed by all architects, engineers and designers, that get involved in any kind of activities on catholic buildings of worship in Italy. In particular two of those documents are particularly interesting in order to understand the aims of CEI with reference to both the construction and the adaptation of buildings of worship:

- pastoral note for the adaptation of churches according to the liturgical reform (Commissione episcopale 1996);
- pastoral note for the design of new churches (Commissione episcopale 1993).

Indeed, the main objectives, needs and constraints of CEI with reference to construction and restoration of buildings of worship are embodied in these documents.

Then, according to the contents of the two pastoral notes, and to interviews with personnel of the CEI's office for building of worship, the main needs of CEI for the operation and maintenance of new churches were defined for estimating the degree of importance of requirements for the cases study.

Although all requirements are important to some extent, both the contents of the two pastoral notes and interviews showed that a prominent part of the needs of CEI for the operation and management of churches are related to elements of closures. Indeed the main concern of CEI was on aspect related to visible degradations on elements (such as walls floors and roofs), heat dispersion, ventilation and safety in case of fire.

---

[1] Servizio Nazionale per l'edilizia di culto.

For these reason the applications were conducted on the most important elements in relation to CEI's objectives:

- vertical perimeter walls,
- vertical external frames,
- ground floors,
- roofs,
- horizontal external walls.

The degrees of importance normally represent the combination of both legal obligations and needs, desire and constrains of clients. In this particular case, as the buildings of worship are subjects to particular conditions for most of the legal obligations, the degrees of importance of requirements have been derived directly from the interpretation of CEI's needs.

First a degree of importance was given to all technological requirements involved with the elements of closure mentioned above. Then a degree of importance was attributed to all user requirements that result in some importance for the CEI according to the pastoral notes and the interviews. Those requirements are:

- Evacuation in case of emergencies,
- Protection against falling,
- Security against intrusion,
- Control of internal air temperature,
- Summer comfort,
- Control of noise,
- Control of ventilation,
- Control of odours.

6.2 Case Studies: Criteria of Selection and Application of the Dashboard for Risk Management

In order to test the risk assessment process, two churches were kindly provided by CEI. Case studies were chosen according to some criteria, such as: position, age and techno-typological characteristics.

Churches of 5–8 years old were chosen. This range was selected because of two reasons: by the one side because CEI gather design documents on digital support only from 2005, and in order to easily allow the exchange of files with designers and bishoprics it was much better to treat projects whose documentation were digital, rather than on paper. Therefore all projects after 2005 were chosen. And by the other side, because ideally it is better to evaluate buildings already constructed, or almost finished, rather than still to be done (because seeing the actual building helps in understanding the goodness of the evaluations carried out). For these reasons cases were chosen that were designed as close as possible to 2005.

Another key driver on which case studies were chosen is their formal and techno-typological characteristics. Indeed, buildings of worship are often characterized by some peculiar features that can create problems in the phase of operation and maintenance. Those features are, for example:

- great heights;
- large interfaces;
- large exposed surfaces;
- areas differentiated for usury and deterioration;
- base of the buildings difficult to reach with vehicles;
- morphology of roofs;
- large paraments that cover elements behind;
- uniqueness of the components, combinations of materials that are not compatible;
- large and numerous skylights, articulation and complexity of the surfaces.

Churches were selected that showed a number of these features. The reason for this choice was to test the method on buildings that was likely to present some critical points to be found.

In conclusion, two churches where selected in Italy, designed 5–8 years ago and that present some potential critical features for maintenance.

For privacy reasons of both communities and designers involved with the case studies, it will not be reported; neither names, nor recognizable images, or drawings of the churches used as case studies in the presentation of work. The cases study will be named: Church 1 and Church 2.

Once the risks have been identified (The most important ones are outlined in Sect. 6.1), the next steps for risk assessment are risk analysis and risk treatment (according to the scheme of Fig. 5.2).

At the risk analysis step the uncertainty over objectives for the phase of operation and maintenance is evaluated. In particular, in order to do so it is necessary to define both, the need for maintenance of element linked to the requirement, and its own maintainability.

Risk analysis is a crucial operation. Indeed, in order to assess the risk over objectives for the phase of operation and design, the uncertainty of requirements that represent those objectives have to be estimated first. Then for all requirements the level of uncertainty has to be multiplied for its degree of importance in order to state a level of risk.

In order to estimate the degree of uncertainty over long-term objectives on the two cases studies (Church 1 and 2) the following 4 steps are to be followed:

- elements involved with all requirements have to be identified,
- maintainability have to be evaluated for these elements,
- spreadsheets have to be used to check whether or not the maintainability of the element is adequate to keep the building working as required,
- a degree of uncertainty have to be defined as result of the previous steps.

A synthesis of the application on the two churches of the process for risk analysis follows for both technological and user requirements.

6.2.1 Church 1

(a) Technological Requirements
In line with the degree of importance of requirements for the CEI (Sect. 6.1) the technological requirements evaluated on Church 1 were those related to all elements of closures. Church 1 presents 4 specific elements of closures: one type of external wall, one entrance door, one horizontal window and the roof. The elements have the following characteristics:

- The external wall is a masonry in bricks with finished in plaster on both sides;
- The entrance door is a double door with extruded profile in copper alloy;
- The window is a horizontal window with metal frame and glass closure;
- The roof is a pitched roof with structure in reinforced concrete and finish in metal plates;
- The ground floor is a ventilated ground floor in concrete and finished in tiles.

For all of the elements mentioned above levels of maintainability were evaluated, using the proper tool presented in this chapter. Results of all evaluations are these reported in Fig. 6.1.

Door

FACTORS OF MAINTAINABILITY	LEVEL
Accessibility	5
Ergonomics	4
Easy of assembly and disassembly	3
Cleanability	4
Substitutability	4
Portability	1
Visibility	6

External wall

FACTORS OF MAINTAINABILITY	LEVEL
Accessibility	5
Ergonomics	4
Easy of assembly and disassembly	0
Cleanability	3
Substitutability	2
Portability	3
Visibility	6

Roof

FACTORS OF MAINTAINABILITY	LEVEL
Accessibility	6
Ergonomics	4
Easy of assembly and disassembly	1
Cleanability	2
Substitutability	3
Portability	2
Visibility	3

Horizontal window

FACTORS OF MAINTAINABILITY	LEVEL
Accessibility	6
Ergonomics	3
Easy of assembly and disassembly	3
Cleanability	4
Substitutability	4
Portability	1
Visibility	6

Ground floor

FACTORS OF MAINTAINABILITY	LEVEL
Accessibility	4
Ergonomics	5
Easy of assembly and disassembly	1
Cleanability	2
Substitutability	4
Portability	3
Visibility	5

Fig. 6.1 Results of the evaluations of the 5 elements of closures of Church 1

Then, the levels of maintainability assessed for each element were compared to the demand of maintainability of all interventions associated with the specific technological requirements, in order to define a degree of uncertainty of those requirements (as explained in Sect. 5.2). For example regarding the example reported in Fig. 6.11, the degree of uncertainty over the requirement of "not hygroscopic" is very high (level 5) because the number of interventions linked with that requirement and that are difficult to conduct are 3, and all of them are scheduled quite frequently. In 30 years the required interventions that are difficult to perform are 70, which is a considerable amount. The requirements were therefore evaluated highly uncertain.

The uncertainty over all technological requirements of Church 1 (in a range from 1 to 5), were then reported into the matrix of uncertainty where the only element presenting an high degree of uncertainty (level 5) resulted to be the roof.

In particular the most of the problems are related to the requirements of the roof that in order to be guarantee over time need inspections. Requirements of this type for the roof are: "not hygroscopic", "control of roughness "and "repellency" (Fig. 6.2).

This is due to the fact that the only intervention that is "difficult to perform" in the Church 1 is the visual inspections of the roof. Indeed the level of visibility attribute to the roof is 3 and the minimum level required for easily conduct visual inspection is 4. This aspect penalizes all requirements of the roof that involve visual inspection.

Moreover, a technical requirement associated to a medium degree of uncertainty (level 3) in church 1 is the "control of roughness" of the ground floor. The reason for this is that the accessibility of the floor is equal to 4 (see Fig. 6.1), as for working on the ground floor a significant amount of furniture must be removed. Then since some of the interventions on the floor linked to the requirement of "control of roughness" need a level of accessibility equal to 5, the uncertainty over that particular requirement has risen to 3 (medium degree of uncertainty).

(b) User Requirements

In line with the method presented in Sect. 5.2, user requirements are computed through Monte Carlo simulations. Monte Carlo allows estimate outputs of processes that depend on numerous factors by sampling random numbers in accordance to a probability distribution. In order to conduct such simulation for the risk on clients' objectives several steps are required:

- All technological requirements involved with the user requirement are defined;
- The degree of uncertainty of all requirement involved is evaluated.
- Random numbers are extracted to simulate real behaviours of set of elements using the distribution of probability around the 5 levels of uncertainty of all technological requirements.
- A simulated frequency is obtained for the user requirements, that can be converted into degrees of uncertainty, on the basis of the results of simulations.

With reference to user requirements the main degrees of uncertainty are those related to the requirements of "evacuation in case of emergency" and "protection

Classes of technological units	2-Closures

Technological units	2.4-Closures on top

Classes of technical elements	2.4.1-Roofs

Typological configuration of technical elements	2.4.1.- Pitched roof (ordome) with structure in reinforced concrete and finish in metal plates (flat or wavy)

Specific technical element	Roof of church 1

ACTUAL LEVEL OF MAINTAIN ABILITY EVALUATED FOR THE TECHNOLOGICAL ELEMENT

FACTORSOFMAINTAINABILITY	LEVEL
Accessibility	6
Ergonomics	4
Easy of assembly and disassembly	1
Cleanability	2
Substitutability	3
Portability	2
Visibility	3

TECHNOLOGICAL REQUIREMENT

Code	Requirements	Uncertainty
TR2	Not hygroscopic	5
TR6	Sound absorption	1
TR7	Light absorption	1
TR12	Control of the intrinsic energy contents	1
TR13	Control of the solar factor	1
TR20	Control of the thermal inertia	1
TR23	Control of roughness	5
TR28	Control of heat loss for transfer	1
TR33	Repellency	5
TR37	Sound proofing	1
TR39	Thermal insulation	1
TR43	Reaction to fire	1
TR46	Resistance to aggressive substances	1
TR48	Resistance to fire	1
TR49	Resistance to frost	1
TR51	Resistance to radiation	1
TR52	Mechanical resistance	1
TR56	Stability chemical reactive	1
TR57	Resistance to shock and impact	1

Fig. 6.2 First page of the spreadsheet to evaluate the degree of uncertainty over all technological requirements of the roof of Church 1

against falling", even though none of those two requirements have a particularly high degree of uncertainty. For both of them the uncertainty is equal to 3, which correspond to a medium, but still significant, level. Indeed, if combined with a high degree of importance, a degree of uncertainty of 3 bring the level of risk up to 15/25, which is quite a relevant one (see Fig. 6.3).

The reasons for the level of uncertainty of both these requirements depend on uncertainty over the "control of roughness" of the floor. Indeed this is one of the technological requirements involved with both user requirements mentioned. In Fig. 6.4 a sample from the computation of the uncertainty over "evacuation in case of emergency" is shown.

This particular user requirement involves 7 technological requirements in Church 1: the "resistance to fire" of 5 elements (wall, ground floor, door, roof, window), the "control of roughness" of the floor and the "ease of operation and handling" of the door. All the requirements, except for the "control of roughness" of the floor, are associated to a very little uncertainty over time. Nevertheless, as the "control of roughness" of the floor is a necessary condition for guaranteeing the "evacuation in case of emergency", its uncertainty alone brought the user requirement to a medium degree of uncertainty. In particular simulations with

Fig. 6.3 Matrix of
uncertainty of user
requirements of Church 1

	User requirements							
Building and building areas	UR2.1 Evacuation in case of emergency	UR4.1 Protection against falling	UR5.1 Security again stintrusion	UR6.1 Control of internal air temperature	UR6.2 Summer comfort	UR7.1 Control of noise	UR9.1 Control of ventilation	UR9.2 Control of odors
The whole building	3	3	2	1	1	2	1	1

Monte Carlo methods indicate 3 as the most likely level of uncertainty (45 %) for this requirement, 4 as second (with a likelihood of 30 %) and 2 as third (with a likelihood of 25 %).

6.2.2 Church 2

Also for Church 2, in order to assess the risk over the requirements presented in Sect. 6.1 the degree of uncertainty over all requirements, both technological and user requirements, were estimated first. Then it was combined with the degree of importance in order to estimate the level of risk of all requirements.

(a) Technological Requirements
Church 2 has 19 specific elements among the type of closures treated in these tests: one type of external wall, 3 external doors, 13 windows, the ground floor and the roof. The elements have the following characteristics:

- The external wall is a masonry in naked concrete on both sides, made of the following layers: (1) reinforced white concrete finished in mechanical hammering, (2) insulating foam panels, (3) vapour barrier and (4) walls in reinforced white concrete.
- External doors are 3: door 1 is door with frame in laminated steel, door 2 and 3 are wooden doors, coated on both sides in oak ribbed brushed lightly bleached.
- External windows are 13: windows 1, 2, 3, 4, 5, 6, 7, 8, 9 and 10 are windows with frame in painted aluminium with thermal break. And windows 11, 12 and 13 are fixed windows in aluminium with thermal break.
- The ground floor is a ventilated floor in concrete with finished in tiles, made of the following layers: (1) paving slabs, (2) substrate of sand, (3) slab of insulation in polyurethane, (4) waterproofing layer, (5) layer of gravel and substrate.
- The roof is a pitched roof with wood structure and finish in metal plates, made of the following layers: (1) roof covering slab of zinc and titanium, (2)

	External wall	Entrance door	Ground floor	Roof	Horizontal window	Ground floor	Entrance door	OUTPUT OF	OUTPUT OF
	Resistance to fire	Resistance to fire	Resistance to fire	Resistance to fire	Resistance to fire	Control of roughness	Ease of operation and handling	SIMULATION [n]	SIMULATION [%]
Level 1	75	75	75	75	75	0	70	0	0
Level 2	100	100	100	100	100	25	100	251	25
Level 3	100	100	100	100	100	75	100	450	45
Level 4	100	100	100	100	100	100	100	299	30
Level 5	100	100	100	100	100	100	100	0	0
								1000	
Random number	33	20	49	23	57	84	15	4	
Level of uncertainty	1	1	1	1	1	4	1		
Random number	53	57	2	16	91	77	38	4	
Level of uncertainty	1	1	1	1	2	4	1		
Random number	23	23	60	48	97	53	57	3	
Level of uncertainty	1	1	1	1	2	3	1		
Random number	25	36	75	50	10	100	38	4	
Level of uncertainty	1	1	2	1	1	4	1		
Random number	87	80	4	35	93	4	16	2	
Level of uncertainty	2	2	1	1	2	2	1		
Random number	96	8	91	22	38	93	81	4	
Level of uncertainty	2	1	2	1	1	4	2		

Fig. 6.4 Degree of uncertainty of the user requirement of "evacuation in case of emergency" for Church 1

waterproof membrane, (3) planking, (4) space for ventilation, (5) insulating layer of extruded polystyrene, (6) vapour barrier, (7) purlins, (8) wooden beams.

For all of the elements mentioned above a level of maintenance was evaluated with the proper tool and results are reported in Fig. 6.5.

Then, using the levels of maintainability assessed for each element, a degree of uncertainty was defined for all requirements, and reported into the matrix of uncertainty.

The result was that elements that present significant degrees of uncertainty over time in church 2 are the roof and all the 13 windows.

Reasons for this are that the accessibility of the roof (level 4) doesn't meet the required conditions for conducting some interventions, such as: inspections and painting. Therefore the propensity of requirements involved with these interventions to keep adequate performances over time results less probable than that of other requirements.

Regarding the windows, the main problems were related to the requirements dependent on interventions that involves disassembly and reassembly the components (like "registration of moving parts"). Requirements of this type on windows

External wall

FACTORS OF MAINTAINABILITY	LEVEL
Accessibility	5
Ergonomics	4
Easy of assembly and disassembly	0
Cleanability	1
Substitutability	2
Portability	3
Visibility	6

Door 1

FACTORS OF MAINTAINABILITY	LEVEL
Accessibility	5
Ergonomics	4
Easy of assembly and disassembly	3
Cleanability	4
Substitutability	4
Portability	2
Visibility	6

Windows 1, 2, 3, 4, 5, 6, 7, 8, 9 and 10

FACTORS OF MAINTAINABILITY	LEVEL
Accessibility	3
Ergonomics	4
Easy of assembly and disassembly	3
Cleanability	4
Substitutability	4
Portability	2
Visibility	6

Door 2 and 3

FACTORS OF MAINTAINABILITY	LEVEL
Accessibility	7
Ergonomics	4
Easy of assembly and disassembly	3
Cleanability	4
Substitutability	4
Portability	2
Visibility	6

Windows 11, 12 and 13

FACTORS OF MAINTAINABILITY	LEVEL
Accessibility	5
Ergonomics	4
Easy of assembly and disassembly	3
Cleanability	4
Substitutability	4
Portability	2
Visibility	6

Ground floor

FACTORS OF MAINTAINABILITY	LEVEL
Accessibility	4
Ergonomics	5
Easy of assembly and disassembly	1
Cleanability	3
Substitutability	4
Portability	4
Visibility	5

Roof

FACTORS OF MAINTAINABILITY	LEVEL
Accessibility	4
Ergonomics	4
Easy of assembly and disassembly	2
Cleanability	3
Substitutability	2
Portability	2
Visibility	4

Fig. 6.5 Results of the evaluations of the 19 elements of closures of Church 2

Classesoftechnologicalunits	2 - Closures
Technologicalunits	2.4 - Closures on top
Classesoftechnicalelements	2.4.2 - Horizontal extern alframes
Typologicalconfigurationoftechnicalelements	2.4.2.a- Iron frame with glass closures
Specifictechnicalelement	F1,F2,F3,F4,F5,F6,F8,F9,F10

ACTUAL LEVEL OF MAINTAIN ABILITY EVALUATED FOR THE TECHNOLOGICAL ELEMENT

FACTORSOFMAINTAINABILITY	LEVEL
Accessibility	3
Ergonomics	4
Easyofassemblyanddisassembly	3
Cleanability	4
Substitutability	4
Portability	2
Visibility	6

TECHNOLOGICAL REQUIREMENT

Code	Requirements	Uncertainty
TR6	Sound absorption	1
TR7	Light absorption	1
TR10	Ease of operation and handling	3
TR11	Easiness of maneuvers	3
TR12	Control of the intrinsic energy contents	1
TR13	Control of the solar factor	1
TR14	Control of light factor	1
TR27	Control of heat loss for air changes	1
TR28	Control of heat loss for transfer	1
TR31	Operability	1
TR37	Soundproofing	3
TR39	Thermal insulation	3
TR43	Reaction to fire	1
TR46	Resistance to aggressive substances	1
TR47	Resistance to biological attack	1
TR48	Resistance to fire	1
TR49	Resistance to frost	1
TR50	Resistance to intrusion	1
TR51	Resistance to radiation	1
TR57	Resistance to shock and impact	1
TR58	Airtightness	1
TR59	Watertightness	1
TR62	Resistance to dust	1

Fig. 6.6 First page of the spreadsheet to evaluate the degree of uncertainty over all technological requirements of windows 1, 2, 3, 4, 5, 6, 7, 8, 9 and 10 of Church 2

are: "ease of operation and handling", "easiness of manoeuvres", "soundproofing" and "thermal insulation" (Fig. 6.6).

Results of Fig. 6.6 are due to the fact that interventions like "registration of moving parts" are "difficult to perform" in all windows of Church 2. Indeed windows in this particular building are craftsmanship elements difficult to be regulated. For this reason the level of maintainability with reference to the factor of "easiness of assembly and disassembly" is 3 for all windows, while the minimum level required for easily regulate moving parts is 4. This aspect penalizes all requirements of the roof involved with the assembling and disassembling of windows or part of windows.

Fig. 6.7 Matrix of uncertainty of user requirements of Church 2

Building and building areas	UR2.1 Evacuation in case of emergency	UR4.1 Protection against falling	UR5.1 Security against intrusion	UR6.1 Control of internal air temperature	UR6.2 Summer comfort	UR7.1 Control of noise	UR9.1 Control of ventilation	UR9.2 Control of odors
The whole building	2	2	2	2	1	1	4	1

(b) User Requirements

With reference to user requirements the main degree of uncertainty in Church 2 is certainly that of "control of ventilation". Indeed the uncertainty on the adequacy of building performance over time with reference to this requirement it was evaluated as high as 4 (see Fig. 6.7).

The reason this user requirement resulted in uncertainty, is that the ventilation of the building strongly depends on the possibility to easily open and close windows and doors when needed. Therefore since the technological requirement of "ease of operation and handling" is quite uncertain (level 3) for all 13 windows of church 2, then the overall "control of ventilation" result is highly uncertain. In particular simulations with Monte Carlo methods (Fig. 6.8) point 4 as the most likely requirement (with a likelihood of 66 %) and point 3 as second (with a likelihood of 34 %).

6.3 Results of Applications

The application of the procedure for risk assessment in architectural design on the two buildings of worship has resulted in the attribution of a level of risk, within a range from 1 to 25, to all requirements that express the main objective of CEI.

The level of risk of all requirements was estimated by multiplying the degree of uncertainty and the degree of importance of all requirements. Then all risks were reported within the dashboard of risks (Figs. 6.9, 6.10 and 6.11).

Risks on the tableau de boards show that in both churches the roof is a critical element. This doesn't surprise since, as highlighted at the beginning of this chapter, some of the peculiar characteristics of the buildings of worship that make maintenance of churches difficult involves the top closure: great heights and the complex morphology of roofs.

	Door 1	Door 2	Door 3	Window 1	Window 2	Window 3	Window 4	Window 5	Window 6	Window 7	Window 8	Window 9	Window 10	Window 11	Window 12	Window 13	OUTPUT OF SIMULATION [n]	OUTPUT OF SIMULATION [%]
	Ease of operation and handling	Ease of operation and handling	Ease of operation and handling	Ease of operation and handling	Ease of operation and handling	Ease of operation and handling	Ease of operation and handling	Ease of operation and handling	Ease of operation and handling	Ease of operation and handling	Ease of operation and handling	Ease of operation and handling	Ease of operation and handling	Ease of operation and handling	Ease of operation and handling	Ease of operation and handling		
Level 1	70	75	75	0	0	0	0	0	0	0	0	0	0	0	0	0	1000	0
Level 2	100	100	100	25	25	25	25	25	25	25	25	25	25	25	25	25	0	0
Level 3	100	100	100	75	75	75	75	75	75	75	75	75	75	75	75	75	339	34
Level 4	100	100	100	100	100	100	100	100	100	100	100	100	100	100	100	100	660	66
Level 5	100	100	100	100	100	100	100	100	100	100	100	100	100	100	100	100	0	0
Random number	55	44	74	42	14	73	78	29	2	56	66	17	59	47	22	60		
Level of uncertainty	1	1	1	3	2	3	4	3	2	3	3	2	3	3	2	3	3	
Random number	7	3	60	1	71	21	25	91	6	93	42	64	63	74	13	69		
Level of uncertainty	1	3	1	2	3	3	2	3	2	4	3	3	3	3	2	3	3	
Random number	83	30	12	40	59	43	73	87	55	58	18	39	23	85	16	96		
Level of uncertainty	1	1	1	3	3	3	3	4	3	3	2	3	2	4	2	4	4	
Random number	54	8	53	55	9	92	16	78	42	0	94	26	75	40	59	4		
Level of uncertainty	1	1	1	3	2	4	2	4	3	2	4	3	4	3	3	4	4	
Random number	42	87	19	63	72	49	67	61	68	78	7	97	75	67	5	12		
Level of uncertainty	1	2	1	3	3	3	3	3	3	4	2	4	4	3	2	2	4	
Random number	78	5	30	45	80	9	94	49	94	46	80	55	12	94	57	65		
Level of uncertainty	2	1	1	3	4	2	4	3	4	3	4	3	2	4	3	4	4	
Random number	92	74	76	91	77	65	69	71	96	39	24	97	51	94	75	74		
Level of uncertainty	2	1	2	4	4	3	3	3	4	3	2	4	3	4	4	3	4	
Random number	92	81	77	100	32	23	21	20	49	76	77	20	33	13	31	55		
Level of uncertainty	2	2	2	4	3	2	2	2	3	4	4	2	3	2	3	3	4	
Random number	64	51	21	4	85	87	100	35	70	9	61	41	3	93	76	48		
Level of uncertainty	1	1	1	2	4	4	4	3	3	2	3	3	3	4	4	3	4	
Random number	83	49	59	34	47	19	57	95	83	57	56	26	2	94	93	71		
Level of uncertainty	2	1	1	3	3	2	3	4	4	3	3	2	2	4	4	3	4	
Random number	39	55	38	24	95	13	22	25	19	93	13	80	46	68	20	19		
Level of uncertainty	1	1	1	2	4	2	2	3	2	4	2	4	3	3	2	2	4	
Random number	84	79	15	67	43	38	76	60	95	50	6	74	48	1	20	7		
Level of uncertainty	2	2	1	3	3	3	4	3	4	3	2	4	3	1	2	2	3	
Random number	1	27	32	78	95	78	81	71	10	66	97	57	67	17	23	55		
Level of uncertainty	1	1	1	4	4	4	4	3	2	3	4	3	3	2	2	2	4	
Random number	75	60	59	6	24	62	18	33	60	42	96	17	85	93	39	10		
Level of uncertainty	2	1	1	2	2	3	2	3	3	3	4	2	4	4	3	2	4	

Fig. 6.8 Degree of uncertainty of the user requirement of "control of ventilation" for Church 2

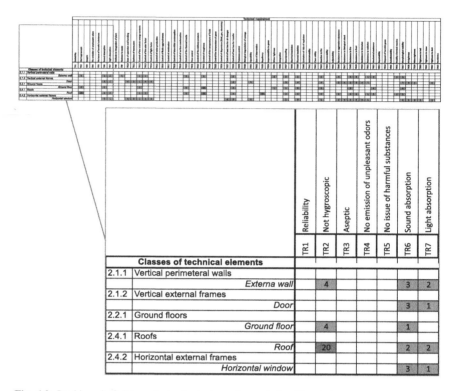

Fig. 6.9 Dashboard of risk on technological requirements for Church 1

Moreover other results that the dashboard presents are the following:

There is a quite relevant level of risks related to the "control of roughness" of the ground floor of Church 1 and to the "ease of operation and handling", "easiness of manoeuvres" and "thermal insulation" of the windows of Church 2. Moreover also the user requirements of "evacuation in case of emergency" and "protection against falling" in Church 1, and "control of ventilation" in Church 2 resulted to be significantly risky.

With reference to the risks pointed in the dashboards it is important to underline the fact that each requirement can be tracked down to the sheet used to evaluate its degree of uncertainty, in order to understand the reasons of its level of risk and, eventually, take action on them.

For example, the risk level of 12/25 for the requirement of "ease of operation and handling" of the window 1 (Fig. 6.12) is determined by the difficulty in assembly and disassembly of the components of the window. In order to reduce such a risk actions can be taken to improve this property of the elements. For example the window could be replaced with another one with a handling system easier to disassembly. Or, alternatively, risk can also be accepted.

Finally, following up with the second part of the risk management process proposed in this chapter, information from future interventions of maintenances on the two churches can be gathered over time, by the one side in order to verify that

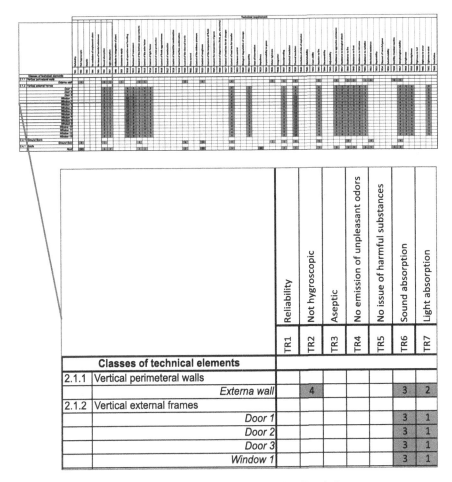

Fig. 6.10 Dashboard of risk on technological requirements for Church 2

	User requirements										User requirements							
	Evacuation in case of emergency	Protection against falling	Security against intrusion	Control of internal air temperature	Summer comfort	Control of noise	Control of ventilation	Control of odors			Evacuation in case of emergency	Protection against falling	Security against intrusion	Control of internal air temperature	Summer comfort	Control of noise	Control of ventilation	Control of odors
Building and building areas	UR2.1	UR4.1	UR5.1	UR6.1	UR6.2	UR7.1	UR9.1	UR9.2		**Building and building areas**	UR2.1	UR4.1	UR5.1	UR6.1	UR6.2	UR7.1	UR9.1	UR9.2
Church 1	15	15	8	4	3	6	4	3		Church2	10	10	8	8	3	3	16	3

Fig. 6.11 Dashboard of risk on technological requirements for Churches 1 and 2

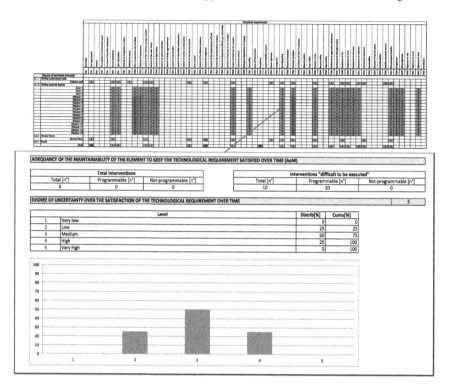

Fig. 6.12 Example of traceability of a level of risk to its causes: the "ease of operation and handling" of the window 1 in building 2

risk assessment in the design was correct and, eventually, to correct them while the process is ongoing and, on the other side, to improve information on databases for future applications (learning by using).

In particular, information that can be gathered from use and maintenance of churches are: time and costs of interventions along with the record of any problem recorded from personnel of maintenance and cleaning companies.

References

Websites and documents available online

Commissione episcopale per la liturgia della Conferenza Episcopale Italiana, La progettazione di nuove chiese. Nota pastorale. Roma, 1993. http://www.chiesacattolica.it/documenti/2002/10/00005808_la_progettazione_di_nuove_chiese_nota_pas.html. Accessed 19 Nov 2012

Commissione episcopale per la liturgia della Conferenza Episcopale Italiana, L'adeguamento delle chiese secondo la riforma liturgica. Nota pastorale. Roma, 1996. http://www.chiesacattolica.it/documenti/2007/03/00012536_l_adeguamento_delle_chiese_secondo_la_rif.html.   Accessed 19 Nov 2012

Servizio Nazionale per l'edilizia di culto. http://www.chiesacattolica.it/snec/00010660_Home_page.html

Chapter 7
Conclusions

The current practice of building management is often characterized by a series of problems that range from discomfort for occupants to high costs of maintenance. There are many cases of old, recent and even very recent architecture designed without adequate control of the technical solutions necessary for their duration, that after a period, literally, fall to pieces. There are many buildings that show that, as time passes, they have been designed without consideration of essential requirements such as stability", "health", "comfort".

The origins of these problems that make building owners and users dissatisfied during the phase of operation and management lie in some critical aspects of the current building process:

- Brief documents are often characterized by a lack of clear references to the client's need for the operation and management phase. This is a potential source of important risk with reference to the long-term performance of buildings. Indeed the way a building can or cannot satisfy a set of needs depends on the degree to which design solutions respond to the needs expressed in the brief document. If clients do not pay enough attention on making explicit requirements for the phase of operation and maintenance, the design may not be adequate.
- Designers and clients have difficulties in defining impacts of design solutions for the future phase of operation and maintenance. An important reason why the propensity of buildings and building components to satisfy their functions over time is not estimated at the design stage is that it difficult to predict. Indeed this propensity of constructions largely depends on two elements that are currently both uncertain at the early stages of the building process: the reliability and the maintainability of buildings and their components.
- The need for operation and maintenance are often given little consideration at the phase of design validation. During the design validation, design proposals are reviewed and their conformity with the brief requirements is checked. However, at this stage it is common to only consider the performance of

© The Author(s) 2015
C. Martani, *Risk Management in Architectural Design*,
PoliMI SpringerBriefs, DOI 10.1007/978-3-319-07449-8_7

buildings with reference to the beginning of the operational phase, while marginal attention is paid to their propensity to maintain this performance through time.

- Processes of learning by using from feedback gathered during the operation and management phase are not common practice in building management. The ability of organizations to capitalize on feedback from the use in order to improve future initiatives relies on whether those organizations reiterate or not the process over time. Even though there are in building construction examples of permanent organizations that promote and manage several building initiatives, generally promoters of construction are temporary multi-organization that do not regularly capitalize on feedback.

The critical aspects mentioned contribute to make the satisfaction of clients' needs in the long term highly uncertain, and therefore risky, because of design choices.

With reference to the design driven risks that affect the operation and management phase, some interpretative key points can be highlighted:

- Objectives for the operation and management are particularly uncertain. This is because the operation and management phase is a long period that starts far in time from the decisional moment of design and, as a matter of fact, uncertainty grows along with the time horizon of previsions.
- Design is the key decisional moment where uncertainty is defined and it requires proper tools and methods for risk management. Aspects of design that largely determine the propensity of buildings to keep performing as required over time are two: the amount of interventions on which performances of elements depend (number and frequency) and the maintainability of elements on which these performances rely. This assertion comes from the fact that the performance of building elements—required to satisfy over time the objectives of clients and users—are very often dependent on maintenance activities. The maintenance dependency on objects' availability over time requires particular attention to be paid to both all the changes they may undergo as a result of the aging effect (its need of maintenance) and on their propensity to be easily maintained or recovered (its maintainability). Thus, the adequacy of the maintainability of elements, compared to the maintainability needed from their interventions, can be taken as an indicator of the uncertainty around the satisfaction over time of the requirements of those elements.
- Feedback from the operation and maintenance phase can serve to improve future designs. As the operation and maintenance phase is the period in time after the design from which the inertial outcomes of the early decisions appear, the feedback from this phase provides vital information. Feedback is precious because it reveals the quality of the decisions taken upstream and brings to light mistakes, forgetfulness or shortcomings in planning and designing, and in this way it helps parties involved with the process to learn from past experiences and do better in the future.

Following up with these interpretative keys the present work proposes an approach for putting in correlation, through the methodology contributed from the studies on risk management, the decisions taken at the brief and design phases of the building process with their consequences on the operation and management phase. The scope of the approach is to support the activities of both instructing and validating the design with reference to the risks in long-term goals. To this purpose a dashboard of risks is introduced where the level of risk are estimated as a result as the combination of the degree of uncertainty and the degree of importance.

The degree of importance is assigned in accordance with clients' needs to all requirements (that express clients' needs), either technological or user ones. The degree of uncertainty is defined through a set of tools and methods created specifically for estimating both the propensity of requirements to be maintained acceptable over time—by combining required maintenance and maintainability of elements—and the level of maintainability of all pertinent elements. Moreover a system for feedback gathering from building use and maintenance has also been provided.

In order to create the body of tools and the methods described, a number of databases was first created with key information, such as: types and frequencies of interventions, maintainability required from all interventions, technological requirements and user requirements.

Finally the approach was tested on buildings of worship, kindly provided by the Italian Council of Bishops (CEI), with well-known problems in operation, where the applications of the dashboard of risks to the design documents indicated a good effectiveness of tools and methods to identify and assess main risks. Indeed, experiment revealed that in the two churches there is a quite relevant level of risk related to the technological requirements of: control of roughness of the floor, ease of operation and handling and ease of manoeuvres of windows, and of thermal insulation of walls, as well as to the user requirements of: evacuation in case of emergency, protection against falling and control of ventilation. These aspects are due to real maintainability problems of some elements.

From the results of the applications, the methodological approach seems to support the hypothesis to use the techniques of risk management as adequate methods to improve:

- The consciousness of designers of the long-term risks in their choices;
- The ability of clients to properly instruct and evaluate design proposals.

Three possible areas of improvement of the dashboard that can be outlined are:

- The feedback gathering system. Feedback from management in the present procedure only provides little information (time and costs of conductions) and it can potentially be expanded to include information on the state of element and spaces.
- The quality of evaluations based on the quality of information included in the databases. Databases created to make evaluations are open to continuous update, and therefore all new projects managed will increase the body of information

available for future evaluations. In line with this assumption the two case studies analysed in the thesis have already contributed to improving the quality of initial databases.

- Elements considered in the method for evaluating uncertainty. Uncertainty evaluations could also consider forecast functional changes of buildings during their lifetime, such as: changing needs and requirements, differences between the functional and the technical service lives of building components, and differences between the service lives of building components and the functional/economical service lives of buildings.

The main characteristic of the dashboard of risks is that as a predictive tool that puts in correlation decisions with effects in the long term, it is a suitable tool for a variety of areas of applications Possible other areas of applications are any kind of process characterized by decisional moments that have a strong impact on the quality of the future phase of use and maintenance. Indeed all contexts with these characteristics would benefit from the propensity of the tableau de bord to link decisions to their impacts in use. Examples of these cases are design-based processes with strong objectives in the long term and the need of proper conditions to maintain these objectives such as design of infrastructures and design of plants.

Printed in the United States
By Bookmasters